T0221527

# GENETICS
## THE SCIENCE OF LIFE

# DNA
# and
# Genes

SUSAN SCHAFER

**Routledge**
Taylor & Francis Group

LONDON AND NEW YORK

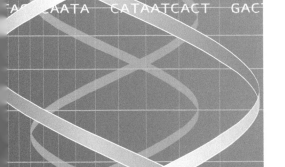

First published 2009 by M.E. Sharpe

Published 2015 by Routledge
2 Park Square, Milton Park, Abingdon, Oxon OX14 4RN
711 Third Avenue, New York, NY 10017, USA

*Routledge is an imprint of the Taylor & Francis Group, an informa business*

Library of Congress Cataloging-in-Publication

Schafer, Susan.
    DNA and genes / Susan Schafer.
      p. cm. — (Genetics: the science of life)
Includes bibliographical references and index.
ISBN 978-0-7656-8135-5 (hardcover : alk. paper)
1. DNA—Juvenile literature. 2. Genes—Juvenile literature. I. Title.
QP624.S33 2009
572.8'633—dc22

                                        2008007918

Editor: Peter Mavrikis
Production Manager: Henrietta Toth
Editorial Assistant and Photo Research: Alison Morretta
Program Coordinator: Cathy Prisco
Design: Patrice Sheridan
Line Art: FoxBytes

ISBN 13: 978-0-7656-8307-6 (pbk)

# Contents

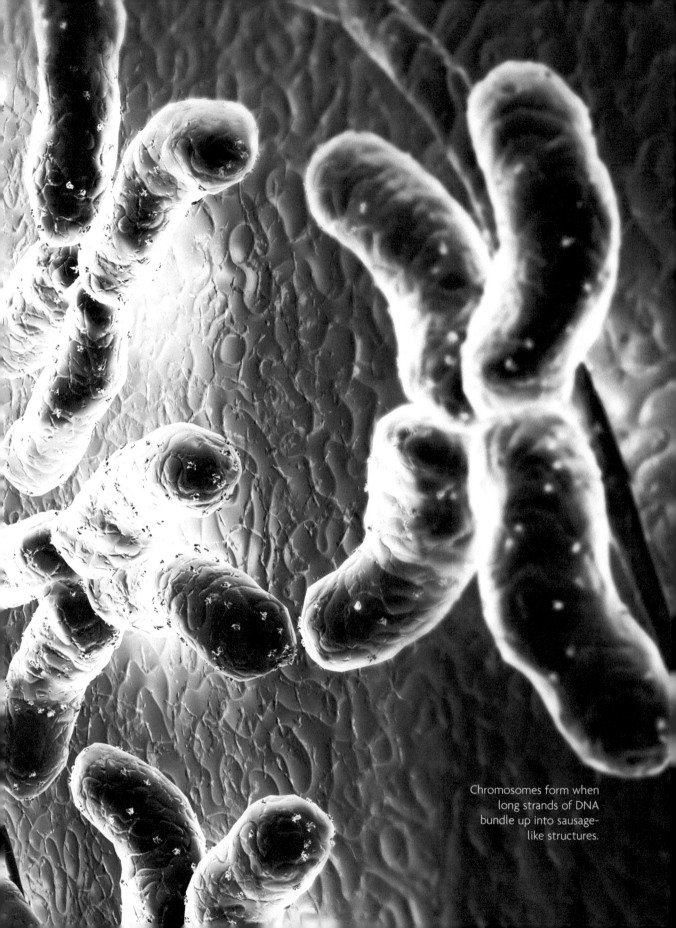

Chromosomes form when
long strands of DNA
bundle up into sausage-
like structures.

# CHAPTER 1

# You Are Your DNA

It all comes down to forty-six microscopic strings of a special chemical called DNA. Without DNA, the cells in your body would not have the instructions they need to survive. It would be as if you were lost in a jungle with no one to tell you where to go. Forty-six complex strands of DNA inside each of a hundred trillion cells work together like a coordinated machine to make what is called a human.

Unless you are an identical twin, your DNA is unique. It controls everything that happens in your body. It determines the person you become. You are your DNA. When you understand DNA, you understand yourself. Like any other science subject, learning about DNA is like learning to speak a new language. The more scientists discover, the more they use new words to describe what they find. So prepare yourself to learn the language of DNA.

## THE CODE FOR ALL LIFE

DNA, which is short for **deoxyribonucleic acid** (*dee ahk see rye boh noo KLAY ik*), holds the codes for life. It is found in the cells of every living

5

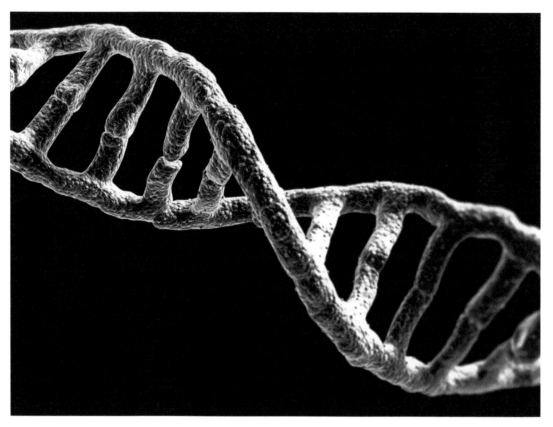

A single strand of DNA forms a double helix, which looks like a twisted ladder.

thing. Like a teacher who controls the lessons in a classroom, DNA is like an instruction manual that controls almost everything a cell does. It is responsible for **heredity,** which is the passing of traits from parents to off-spring. Because DNA is passed from one **generation** to the next, it is entirely possible that you could have some of the same exact DNA that an ancestor had hundreds of years ago. People pass away, but in a way, DNA lives on.

DNA is held within a container called the cell. The cell is the basic unit of life. It is the smallest living thing. Your body is like a city teeming with trillions of individual cells. Cells cling together to hold you together. As the people work together to help their city or town prosper, your cells work as a team to keep you alive.

Inside each cell are smaller structures, called **organelles,** or "little organs." Each organelle has its own special job. One of these organelles, the

## DOWNLOAD

- The initials *DNA* stand for deoxyribonucleic acid.

- DNA is found in the cells of all living things.

- The twisted ladder shape of a DNA molecule is called a **double helix.**

- Instructions from the genes on DNA determine the characteristics of all living things.

- DNA is passed on through heredity.

**nucleus,** has the job of holding the DNA, so it is often called the control center or "brain" of the cell. The control center can be compared to a control tower at an airport. The air traffic controller guiding the planes from inside the tower is like the DNA. The word *nucleus* comes from the Latin word for "kernel" because it looks like a dark, round nut when viewed under low magnification on a microscope.

Inside the nucleus, the DNA works like a big boss, sending out instructions for the cell to make **proteins.** Proteins perform various tasks in the body, from building new organelles to speeding up chemical reactions. Only when a cell reproduces does the nucleus break down and free the DNA. At that point, the DNA has copied itself, so

**ORGANELLES**

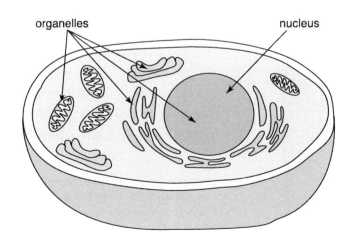

organelles                                          nucleus

Like people in a factory, organelles work together to keep a cell running smoothly. The nucleus is the organelle that holds the DNA.

**PROKARYOTIC CELLS**

chromosome
(no nucleus)

Bacterium

**EUKARYOTIC CELLS**

nucleus

Plant cell

Animal cell

Unlike eukaryotic cells, prokaryotic cells do not have membrane-covered organelles, so their DNA is not contained within a nucleus.

there is twice as much as normal. When one cell divides into two, each new cell has an exact copy of the original DNA. This copying is important, because if it did not occur the new cells would not have the necessary instructions, and they would die. Without new cells to replace old cells, your body would not last long.

You are a **eukaryote** *(yoo KAR ee oht)*, along with other plants and animals, which is just a fancy way of saying that your cells have a "true nucleus." The other type of living cell is a **prokaryote** *(proh KAR ee oht)*, which means before the nucleus. Prokaryotes are primitive cells that lived on Earth long before the eukaryotes. Prokaryotes do not have a nucleus, and their DNA simply floats inside the cell. **Bacteria** are prokaryotes.

The nucleus is not the only place eukaryotes have DNA. Organelles called **mitochondria**

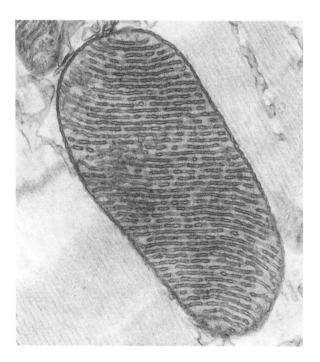

Looking like a fingerprint, a mitochondrion contains many folded membranes that provide a large surface area for producing energy in a cell.

(*my tuh KAHN dree uh*; *singular,* mitochondrion) are like bacteria. They have DNA, but not inside a nucleus. Mitochondrial DNA uses a slightly different code for sending messages. So you actually have two different kinds of DNA in your body!

Because the DNA is different in mitochondria, some scientists believe that they were once prokaryotes that were swallowed by eukaryotes. For some reason, they were not digested. The prokaryote thrived inside the eukaryote because it was protected. The eukaryote benefited because the prokaryote made extra energy that the host eukaryote could use. When the eukaryote multiplied, the prokaryotes were also passed on to future generations. Eventually, they became mitochondria. The mitochondria now provide the energy that you need for your cells to work.

## TAKING A CLOSER LOOK

DNA is so small that it takes a powerful electron microscope to see it. **Electrons** are negatively charged particles that orbit around all atoms. They have the ability to flow and create electricity. With an electron microscope, scientists shoot electrons at objects that cannot be seen in any other way. This process can magnify things up to 2 million times. The compound light microscopes that might be used in a science classroom only magnify up to ten, forty, or a hundred times.

As the electrons bounce off an object or travel through it, depending on the electron microscope, they are detected by special equipment. The equipment then sends an image to a monitor or computer. Scientists have spent many years improving tools that allow them to take a closer look at DNA **molecules.**

Look at a strand of your hair. The width of a DNA strand is about 40,000 times thinner! The length of a DNA strand varies, depending on how many messages it carries, but if all of the DNA from just one cell was laid end to end, it would stretch more than 2 meters (about 6 feet). If you took the DNA out of all the cells in an average adult human—about a 100 trillion cells—it would wrap around the earth's equator nearly 5 million times. And you thought running a mile in gym class was a long way!

Electron microscopes, such as the one seen on the left, allow scientists to take detailed pictures of DNA and other super-microscopic objects—those objects that are so small they cannot be seen with a regular microscope. The nuclear chromosome shown on the right is an example of such an image.

The search for DNA starts with the cell, which spends most of its time in a stage of growth called **interphase.** At this time, the cell is not reproducing or dividing. During interphase, the nucleus contains many coils and loops of a material called **chromatin.** The chromatin is about half DNA and half protein.

### "BEADS-ON-A-STRING" FIBER

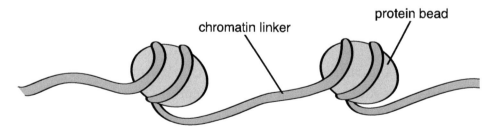

Chromatin, which will coil to form chromosomes during cell reproduction, is made of DNA intermittently wrapped around small beads of protein. A protein bead with its linker strand of DNA is called a nucleosome.

If you pulled out the coils and loops of chromatin, you would see that the DNA wraps around bead-like clusters of proteins. The DNA goes around one protein bead about one and a half times and is then connected to the next bead by a short stretch of DNA called linker DNA. Scientists call this structure, which looks like a necklace, a "beads-on-a-string" fiber. One bead with its linker DNA is called a **nucleosome.**

If you popped the protein bead out of a nucleosome, you would be left with pure DNA, which is made of two strands of chemicals called **nucleic acids.** (That's where the *NA* in DNA comes from.) The two nucleic acids are held together in the middle by hydrogen bonds. You might recognize the

**POP-UP**

When a cell reproduces, the strands of looped chromatin bend, fold, and coil into rod-like structures called **chromosomes**. The **chromatin** in a chromosome is packed up to 10,000 times tighter than when it is in **interphase**. Without packing, the chromosomes might get tangled or break when they move during cell division. It would be like having 46 very long pieces of string bunched together. It would be easier to separate them if each one was rolled into a ball first. Human cells have 46 chromosomes, or 46 condensed strands of chromatin (23 from the mother and 23 from the father). Chromosomes get their name from the Greek term for "colored bodies," because they are easily seen under the low magnification of a microscope.

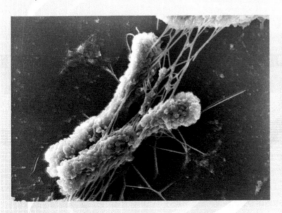
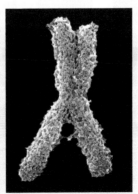

**NUCLEOTIDE**

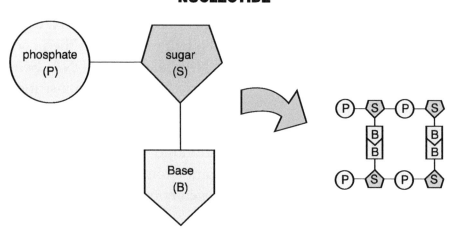

A nucleotide is formed when a phosphate, a sugar, and a base are bonded together. Strings of nucleotides, bonded together by their bases, form the double helix of DNA.

word *hydrogen* from the formula for a molecule of water, or $H_2O$.  Water is formed when two hydrogen atoms bond to one oxygen atom. Hydrogen bonds are strong enough to hold the nucleic acids together most of the time, but weak enough to be broken when the DNA unzips to send a message.

The two bonded nucleic acids are twisted all the way down the DNA molecule into what is called a **double helix.** A helix is shaped like a spring. Two springs coiled together side by side would form a double helix. Imagine the two tracks of a roller coaster twisted into a spiral like a double helix. What a ride that would be!

**V I D E O   C L I P**

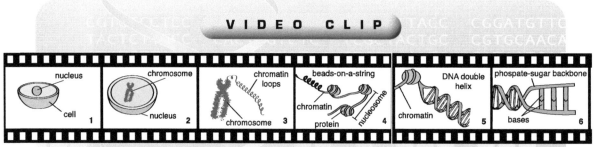

1. The nucleus is found inside of the cell; 2. During cell reproduction, chromosomes form inside the nucleus as the membrane around the nucleus breaks down; 3. A chromosome is made of loops of chromatin; 4. Chromatin is made of DNA and protein; 5. DNA forms a double helix; 6. The double helix is formed by two phosphate-sugar backbones joined in the middle by bases.

A closer look at a nucleic acid would show repeating units called **nucleotides.** A nucleotide is made of a special chemical called a phosphate, which is bonded to a sugar. The sugar is then bonded to a **base.** It is shaped somewhat like a reversed capital letter *L*, with the phosphate on top, the sugar at the corner, and the base at the right side of the foot. The base of one nucleotide bonds with hydrogen to the base of another nucleotide. The two nucleotides form a squared U shape. The nucleotide pairs then bond top and bottom to more nucleotide pairs, forming the railroad track structure of DNA. This structure is twisted to form the double helix.

There are actually two kinds of nucleotides involved in sending messages. Each has a different type of sugar. A nucleotide of DNA contains a deoxyribose sugar. (The *D* in DNA.) The DNA is stationed inside the nucleus and has an assistant carry its messages out to the cell. The assistant is called RNA, or **ribonucleic acid.** A nucleotide of RNA contains a ribose sugar. (The *R* in RNA.)

## THE DNA LANGUAGE

The base on a DNA nucleotide is the foundation for the DNA language. There are only four letters, or bases, in the DNA alphabet. In spite

**DNA TRIPLET**

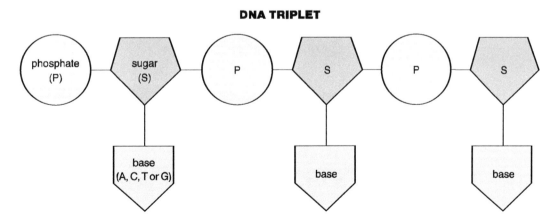

A DNA triplet is made of three bases in a row on one side of a double helix. The bases can be any combination of adenine (A), thymine (T), cytosine (C), or guanine (G). One triplet forms one word in the DNA language.

of this limitation, the bases of DNA can be combined and recombined in enough ways to create messages much longer than any that could be sent in a text message on a cell phone.

The four bases of DNA are chemicals called adenine, thymine, cytosine, and guanine. To make it easy, scientists just use the symbols A, T, C, and G. When the two sides of the double helix bond in the middle, A always bonds with T and C always bonds with G. (An easy way to remember which bases pair together is that the A and T both have straight edges and the C and G are both rounded.) When bonded, the two bases are called base pairs.

### ALERT !

Ultraviolet (UV) light from the sun causes cancer by passing into the nuclei of your skin cells. It carries so much energy that it can break or cause changes, called **mutations,** in DNA. Although cells can repair some damage, unrepaired DNA can cause the cells to multiply out of control and destroy other normal cells. Uncontrolled cell division is called cancer. So remember to wear sunblock whenever you go outside!

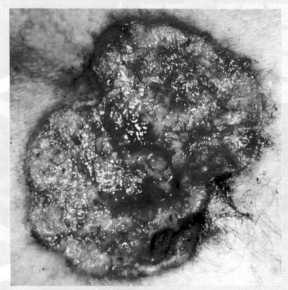

A cancerous lesion on human skin, most commonly caused by overexposure to ultraviolet light from the sun.

The entire DNA from just one human cell contains about 3 billion base pairs, which would be about 1 billion triplets. A **triplet** is made of three bases in a row on one side of a double helix. It would be like a three-letter word. If you had to type one billion three-letter words for an English essay, you would have to type 500 words an hour nonstop for more than 200 years.

A **gene** is formed from hundreds or thousands of triplets running along a DNA strand. Each gene carries its own special message for the cell. If you typed out all of the bases on a single gene using the base symbols, it might look like this: ATGCGAGAATCGGAATTCCCGGTATCGATAACTTGGAATTGA, except that most gene messages are significantly longer. DNA might seem to have no punctuation, but it actually has "start" and "stop" triplets to keep the gene messages separated. In the example, the start triplet is "ATG" and the stop triplet is "TGA."

The number of genes in the DNA of a human cell was once estimated to be around 100,000. It was assumed that an **organism** as complex as a human would need at least that many genes. The latest evidence suggests that humans have about 20,000 to 25,000 genes, as do rats, mice, and even roundworms. Complexity may have more to do with the types of messages genes carry than with the number of genes there are.

Scientists are still trying to pinpoint the exact number of genes in different organisms. Estimates vary, depending on the gene-counting methods that are used. Much of the DNA in a cell has no function, which makes it difficult to study. But with new tools and technologies, scientists continue to make new discoveries on the road to cracking the code of life.

The secrets of DNA and its language of bases have taken scientists many years to uncover.

# CHAPTER 2

# Discovering DNA

Imagine the excitement of discovering the key to life. Many scientists have experienced this feeling in their search for DNA. Like looking for a needle in a haystack, one scientist after another has picked away at the question of what makes each living thing unique. It is likely that researchers will never reach the needle, because each new bit of information leads them in a new direction. For science, there is no end to discovery.

The birth of DNA research began over a hundred years ago with the study of heredity in flowers. The early years of this research were spent untangling the mysteries of exactly what it was that passed on information from generation to generation. Once scientists identified DNA as the culprit, they wanted to know what it looked like and how it worked. The golden years of DNA study still lie ahead. Although controversial, scientists predict that one day people will regularly choose the genes of their children, cheaply clone their favorite pets . . . or even clone themselves!

## THE BIRTH OF GENETICS

**In the mid-1800s,** an Austrian monk named Gregor Mendel tended the garden at his monastery. He was particularly interested in pea plants. Although the peas were important for food, Mendel was more interested in studying them than eating them. He noticed that the pea plants did not always look the same. They were various heights and sometimes the flowers were different colors or grew on different parts of the plant. The pods that held the peas were various shapes and colors. The peas themselves might be green or yellow, smooth or wrinkled.

Mendel started experimenting. First, he crossed pea plants that only produced purple flowers. He did this by using the pollen from a purple flower on one plant to pollinate the female structure in the purple flower of another plant. He found that parents with purple flowers only produced **offspring** with purple flowers.

He also crossed pea plants with white flowers. Two white-flowered parents only produced offspring with white flowers. However, when he crossed a purple-flowered plant (that could only produce purple flowers) with a white-flowered plant (that could only produce white flowers), the resulting plants had purple flowers. No matter how many times he crossed purple with white, he got the same results—all purple. Why did the white flowers disappear?

Mendel then took the purple-flowered plants from the purple-and-white cross and let them self-pollinate. The next generation had mostly purple flowers . . . but there were some white

Gregor Mendel studied the inheritance of traits in pea plants, laying the groundwork for later studies of genetics.

## MENDEL'S PEA EXPERIMENT

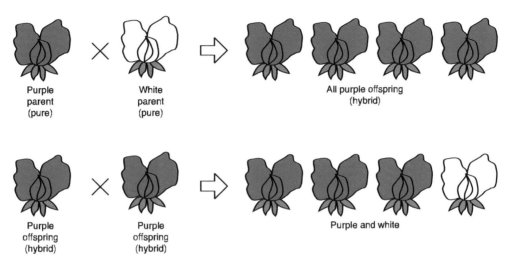

| Purple parent (pure) | White parent (pure) | All purple offspring (hybrid) |

| Purple offspring (hybrid) | Purple offspring (hybrid) | Purple and white |

*Top:* When Mendel crossed pure purple-flowering plants with pure white-flowering plants, all of their offspring were purple. The offspring were not pure purple, but hybrid purple because they carried one "unit" (gene) from each parent, one purple and one white. The white trait was hidden, or recessive, in the hybrid. *Bottom:* When Mendel crossed the purple hybrids with each other, they produced both purple and white offspring. The white trait reappeared in some offspring when two white genes from the hybrid parents were recombined.

flowers! The characteristic for white had reappeared. Mendel believed that the white trait had never been lost, but was hidden somewhere in the purple offspring. He concluded that each **generation** of plants has two "units" of information, one from each of its parents, and that some units were stronger than others. Little did he know that his units would turn out to be the genes on DNA.

Because the purple unit appeared to be stronger than the white, Mendel called it the **dominant trait**. Because the white disappeared, he called it the **recessive trait**. It receded, or withdrew, into the background when it was paired up with the dominant trait. The dominant trait was like a big bully that told the recessive trait to get lost.

Mendel tried the same thing with other pea-plant traits. He discovered that smooth pods were dominant over bumpy pods, yellow peas dominant over green peas, and round peas dominant over wrinkled peas. He found that tall plants were dominant over short plants. These results were to become the foundation for all future studies of heredity. Because of his work, Mendel is now known as the father of genetics.

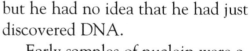

## DOWNLOAD

- In the mid-1800s, Gregor Mendel was the first to study heredity.
- In 1929, Phoebus Levene identified the nucleotide, a phosphate-sugar–based unit on DNA.
- In the 1940s, Erwin Chargoff determined that the amount of adenine in any DNA sample always equaled the amount of thymine, and the amount of cytosine always equaled guanine.
- In 1952, Rosalind Franklin photographed the helical or spiral shape of DNA.
- In 1953, James Watson and Francis Crick built the first model of the DNA double helix.
- Through time, each scientist adds another piece to the puzzle of DNA knowledge.

In the late 1800s, a Swiss doctor named Johann Miescher studied white blood cells that he found in the pus of infected wounds. Pus contains the white blood cells that help to fight disease. Miescher discovered a new chemical inside the nuclei of the white blood cells. He called it nuclein, but he had no idea that he had just discovered DNA.

Early samples of nuclein were a mixture of both RNA and DNA. Eventually Miescher isolated pure DNA, but he still did not know what it was. Then a student showed that DNA was a nucleic acid that was found only in chromosomes.

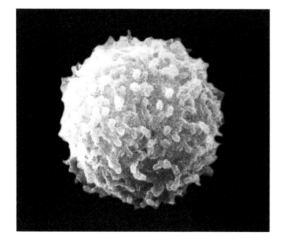

DNA was first isolated from white blood cells found in pus.

Sequencing machines use special lights to identify the four bases of DNA. In one method, the bases are labeled with dyes. A laser light is passed over the labeled bases, which sends a signal to a computer for identification. A newer and even faster method washes the bases, one at a time, across a special chip. When each base finds its complementary base on a DNA strand, it triggers a light signal. The signal is then sent to a computer for analysis.

Unfortunately, both Miescher and his student thought nucleic acids were chemically too simple to hold the information for directing life. Along with other scientists of the time, they believed that proteins, which are long and complex, were responsible for heredity. They did not know that proteins were actually made by instructions from the DNA.

## UNEARTHING THE GENE

In the early 1900s, Danish **botanist** Wilhelm Johannsen reconfirmed Mendel's earlier work, using beans. He also concluded that "units" inside the beans were responsible for passing on traits. Johannsen called these units **genes,** from the Greek word *genos*, which means descent—as in, you are a **descendant** of your ancestors because their genes were passed on to you.

Johannsen also coined the words **genotype** and **phenotype.** The genotype refers simply to the types of genes that an organism has. The phenotype is the physical characteristic that an individual has because of its genotype. Just think *genotype* for *genes* and *phenotype* for *physical*. For example, your genotype has several genes that tell your body what color your hair will be. But your phenotype is the actual color of hair, such as red or black, which you have because of those genes. You cannot see the genes in the genotype, but you can see a phenotype if it is on the outside of the body.

By 1928, an English doctor named Frederick Griffith had demonstrated that bacteria pass on traits through heredity. Most bacteria are harmless, but some can make you sick. Griffith worked with two **strains** of a bacterium that causes pneumonia, a severe lung disease in humans. One strain killed people who were infected. The other strain did not. Griffith boiled the killer strain and injected it into mice. The mice did not die, so he knew he had killed the bacteria. Then he mixed the dead killer strain with the harmless living strain, and the harmless strain became a killer! The killer trait had been passed on to the harmless strain.

Griffith called the transfer of characteristics from one bacterium to the other a "transforming principle." To transform means to change from one thing into another. Griffith was not aware that the transformers were actually the genes that had been described earlier by Johannsen.

Soon after, an American scientist named Phoebus Levene identified the parts of a DNA molecule. He showed that a phosphate group, a deoxyribose sugar, and one of four bases were bonded together in what he called a **nucleotide**. He believed DNA was a string of only four nucleotides in a row. Like other researchers, he assumed such a simple molecule could not possibly hold the genetic code.

As some scientists were working out the **structure** of DNA, others were trying to figure out how genes worked. Some had suggested that a gene works by making one specific protein or **enzyme**. Different enzymes help run the chemical reactions inside a cell. But this idea was ignored until 1941, when George Beadle and Edward Tatum used x-rays and ultraviolet light to create **mutations** in bread molds. Normally, bread molds make their own vitamins, but the mutants created by Beadle and Tatum could not. The mutants only survived if particular vitamins were added to their food. When the scientists studied the genes of the mutants, they found only one gene was different from the original bread mold. This led them to the "one gene–one enzyme" **hypothesis**, meaning that each gene works by making only one particular protein or enzyme.

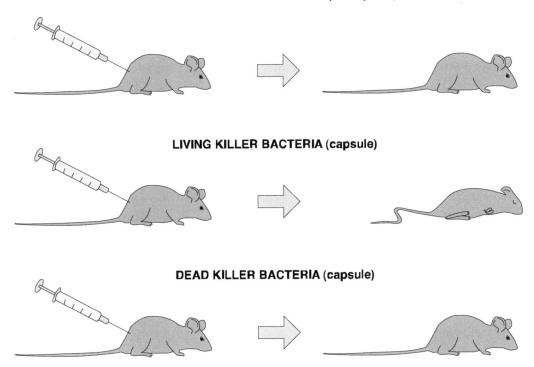

**LIVING HARMLESS BACTERIA (no capsule)**

**LIVING KILLER BACTERIA (capsule)**

**DEAD KILLER BACTERIA (capsule)**

**DEAD KILLER MIXED WITH LIVING HARMLESS (capsule)**

In 1928, Frederick Griffith experimented with two strains of bacteria: a harmless strain that was not covered by a capsule and a killer strain that was (the capsule makes the bacteria deadly). When he examined the bacteria from his last experiment *(bottom)*, he was shocked to see that the harmless living bacteria were covered in a capsule. The harmless bacteria had inherited the deadly trait from the killer bacteria.

At this point, scientists knew there were transforming units or genes. They also knew that DNA was found inside the chromosomes, which are located in the cell's nucleus. In 1944 a team of U.S. scientists led by Oswald Avery put the two ideas together and finally identified DNA, and not proteins, as the molecules of heredity. First, they destroyed various proteins in bacterial cells, but the cells still passed their traits to the next generation. Then they destroyed the DNA, but not the proteins, in the same cells. The bacteria did not pass on their traits. In spite of this, many scientists remained skeptical. Because they did not know that DNA was a long molecule, they could not see how such a simple molecule with only four bases could hold information.

It was Erwin Chargaff of Austria who discovered a pattern of the four bases. He studied the DNA from many different kinds of cells, and always found that the amounts of adenine and thymine were the same. The amounts of guanine and cytosine were also always the same. If he found 20 percent of adenine in a sample, he always found 20 percent of thymine. Of the remaining 60 percent, 30 percent would always be guanine and 30 percent always cytosine. If he found 10 percent adenine, he found 10 percent thymine, 40 percent guanine, and 40 percent cytosine. His discovery came to be known as Chargoff's rule, or A = T and C = G. (Remember that the letters A and T have straight edges so they go together, and the letters C and G are both rounded so they go together.)

**PAIRING OF THE DNA BASES**

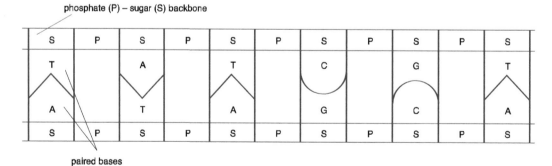

Erwin Chargoff discovered that the amounts of adenine (A) and thymine (T) in a DNA sample are always the same, and that the amounts of cytosine (C) and guanine (G) are always the same. This is because A always pairs with T and C always pairs with G. The above sample contains four adenines, so how many thymines will there be?

For the bases to be in equal amounts in the DNA, scientists realized that they had to be bonded together in pairs. This helped them work out the actual structure of DNA. They also realized that DNA is **species** specific. In other words, different organisms have different amounts of bases in their DNA. This makes sense because a message for creating a turtle's shell would be different from a message for making a frog's skin.

## PURSUIT OF THE DOUBLE HELIX

**In the 1950s,** scientists raced to discover the actual shape and structure of DNA. At the California Institute of Technology, Linus Pauling had discovered that many proteins were in the shape of a helix or spiral. Other scientists suggested that DNA might be shaped similarly, but it took Rosalind Franklin, James Watson, and Francis Crick at Cambridge University in England to prove it.

Franklin worked with **x-ray diffraction** to take pictures of DNA. She wanted scientific evidence about its structure before trying to build a

model. The DNA she used was an **extract** from a cow's thymus. The thymus is a gland just below the base of the neck behind the breastbone that is part of the **immune system**. It contains lots of white blood cells and, therefore, lots of DNA.

Franklin pulled long, elastic strands of DNA from the thymus extract and hung them in front of a camera. Then she sent an x-ray beam through them, which projected an image onto x-ray film.

Rosalind Franklin created the first images of a DNA molecule.

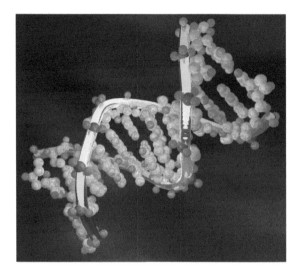

A computerized image of a DNA molecule shows the X, which is formed as the two sides *(blue)* of the double helix twisted around one another.

The resulting images showed that DNA was a long, thin molecule. It certainly was not short, as proposed by Levene thirty years earlier. In 1952, Franklin took the clearest picture of DNA to date; the image now commonly known as Photograph 51. It showed an X-shaped image. DNA was a helix!

In the meantime, Watson and Crick had unsuccessfully been trying to build a model of DNA, using what was known about nucleotides. But they could not figure out how things were hooked together. They struggled with the model for years, until a scientist who had worked in the same lab as Franklin showed them Franklin's photographs. Years before, Franklin had filed the photographs away to work on other projects, and she never knew that they had been removed by her co-worker.

When Watson and Crick saw Photograph 51, they realized that DNA was probably a double helix. They tried another model. Using the sugars

**BLOG**

The scientist who used x-ray diffraction to discover the helical shape of DNA wrote in her notebook:

> Either the structure is a big helix or a smaller helix consisting of several chains.

—Rosalind Franklin, *Journal*, 1952

James Watson, thirty years after he and fellow scientist Francis Crick received the 1962 Nobel Prize for building the first model of DNA.

and phosphates for the sides of the double helix, they paired up Chargaff's **complementary bases** across the middle. It worked. They had built the first model of DNA.

In 1962, Watson and Crick received the Nobel Prize in Physiology or Medicine for their work. In the meantime, Franklin had died of cancer, probably from working unprotected with the very x-rays that gave her Photograph 51. She never knew the part she played in the discovery of the structure of DNA. Watson and Crick's model, with many of the original metal plates and rods, has been rebuilt and now stands in the National Science Museum in London, England.

ALERT

X-rays can cause mutations in DNA. Rosalind Franklin worked with an x-ray beam turned on and without a protective apron. The very x-rays that she used to discover the helical structure of DNA probably caused mutations or changes in her own DNA, leading to the cancer that killed her at age thirty-seven.

## DECODING DNA

**Once** their model was built, Watson and Crick could see that DNA's double helix could be easily "unzipped" down the middle by breaking the bonds between the bases. The bases would then be exposed, allowing a way for new bases to come in and bond with the exposed bases. In this way, DNA could be **replicated**, or copied. Finally, someone had an explanation for how DNA could pass on information. If a cell could make a complete copy of its DNA, then the original cell could keep one copy of the instructions needed to live, and the other copy could be sent to the new cell that was formed.

Watson and Crick also described how proteins could be made from the base messages on the DNA. Instead of making a copy of the entire DNA, part of the double helix could unzip and make a copy of just one gene. The gene copy could then be sent into the cell, where a protein could be made. DNA could both store and pass on messages.

**UNZIPPING THE DNA DOUBLE HELIX**

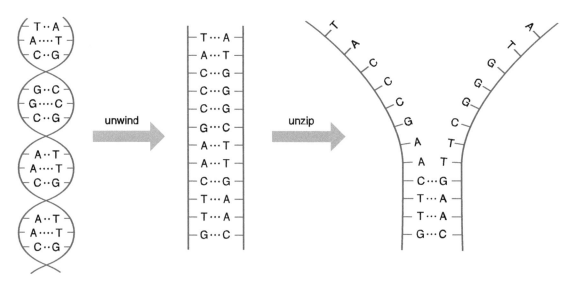

When DNA replicates, or is copied, the double helix unwinds. The bonds that hold the bases together then break, allowing the DNA to unzip. Once the bases are exposed, the DNA is copied.

By the 1960s, a number of scientists figured out how the genetic code worked. In 1961, Crick and another scientist determined that genes worked on a "three-letter word" system. They found that each triplet of bases (three bases in a row) on the DNA, when unzipped and exposed, could form a "word." Each word on the DNA could then be changed into what they called a **codon** on the RNA. Like passenger cars on a train, RNA could carry strings of codons to the cell, forming "sentences" or messages.

Each codon is a complement of the original triplet. For example, the triplet CCG on the DNA forms the codon GGC on the RNA. Each codon signals a specific amino acid. Amino acids are combined to make proteins. Different arrangements of amino acids make different proteins, which are needed for almost all cell activities.

Another team of American scientists, led by Har Gobind Khorana, discovered which amino acids were added to a protein by each of the codons. For example, AGA always tells the cells to make an amino acid called arginine, which means the three nucleotides AGA in that particular order always causes the amino acid arginine to be chemically added to a protein string. GAG, on the other hand, always codes for glutamine. They use the same letters, but in a different order, so they cause different amino acids to be added.

They also discovered that several different codons were "stop codons." Stop codons tell the cell when to stop adding amino acids to a protein. In 1968, Khorana and his team received the Nobel Prize in Medicine for their groundbreaking work.

## FAKE GENES AND THE FUTURE

Khorana's team took their work a step further and created the first **synthetic**, or artificially made, DNA. Scientists can now order "genes" online as easily as you might download a song from iTunes. Today, artificial genes are used to study diseases and change genes in organisms. The characteristics of a living thing can be altered by introducing artificial genes. Scientists have already used genetic engineering to make crops more resistant to insect damage. Perhaps one day you could be engineered so you would be less likely to catch a cold or get the flu.

## RECOMBINANT DNA

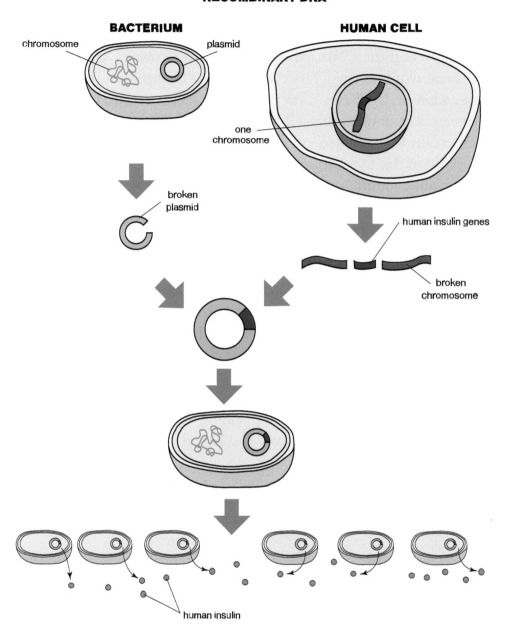

An extra loop of DNA called a plasmid is removed from a bacterium *(top left)*. An enzyme is used to break open the plasmid. The chromosome that carries the insulin gene is removed from a human cell *(top right)*. An enzyme is used to remove the human insulin genes from the chromosome. Recombinant DNA is formed when the plasmid and human insulin genes are stuck together using another enzyme *(middle)*. The recombined plasmid is put back into a bacterium. When the bacterium reproduces, the new cells carry the human insulin genes and, therefore, make human insulin *(bottom)*. The insulin is collected to treat diabetes.

In the last thirty years, scientists have determined the DNA sequences—the order of all the bases on the DNA—of bacteria, fruit flies, a roundworm, mice, and humans. Knowing the sequences tells scientists which genes are present and what proteins the genes make. At first scientists sequenced by hand, but now they use machines that are faster and easy to use. They have determined how to make large quantities of the same gene, which makes it easier to study. They have also identified enzymes that can be used to cut DNA at specific places, so they have smaller pieces to work with. They have also created the first **recombinant DNA**.

Recombinant DNA contains genes that would not be found naturally in a species. For example, scientists have inserted the gene for making human insulin into bacteria. When the bacteria reproduce, they pass on the gene, and more and more bacteria make insulin. Insulin controls the level of sugar in your blood stream. People with diabetes do not make the insulin they need, so they get it from the bacteria. A number of companies around the world now use recombinant DNA technology to make medicines for treating human diseases.

Some day, you may carry a map in your wallet, not of your town or state, but of the genes on your DNA. You might be cured of a disease using recombinant DNA. You might choose the genes of your children. No one knows exactly how advancements in DNA technology will affect people in the future; it is only certain that they will.

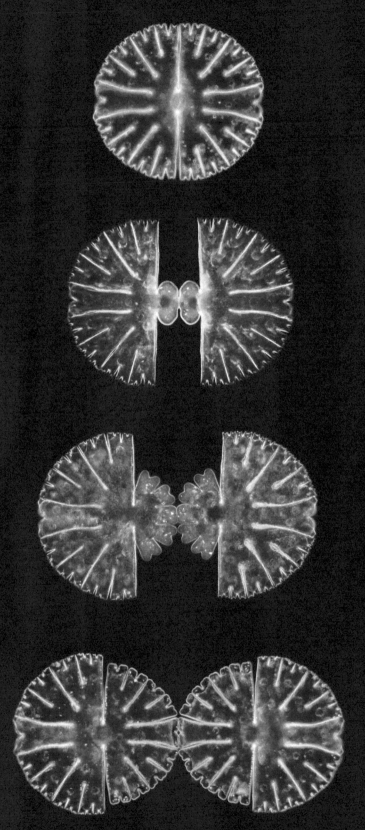

A green alga divides asexually, passing on its DNA to a new generation.

# CHAPTER 3

# DNA Is Life

DNA defines you as a living thing. Without it, your cells—and the cells of every other creature on Earth—would not be able to carry out the processes of life. Every species, including humans, would become extinct. DNA tells cells how to create energy, respond to things in the environment, grow, and reproduce. But what is it that makes something alive?

## THE FORMS OF LIFE

There is one basic form of life—a cell. Every living thing is made of at least one cell. Bacteria are **unicellular**, or single-celled. Most bacteria are so small they cannot be seen with the naked eye. Plants and animals are **multicellular**. They are made of more than one cell. You have around 100 trillion cells, depending on your size. The larger you are, the more cells you have.

But being made of cells with DNA is not what makes you alive. The wood used to make furniture, the leather used to make jackets, and cinnamon sticks all contain cells and DNA, but they are not alive. A living thing

**UNICELLULAR**
**(made of one cell)**

chromosome

nucleus

bacterium
(prokaryote)

protists
(eukaryotes)

**MULTICELLULAR**
**(made of many cells)**

mold

human

plant

(eukaryotes)

Unicellular organisms are usually so small a microscope is needed to see them. Multicellular organisms can grow larger by adding more cells. The cells may be specialized to perform different jobs, making multicellular organisms more complex.

has to reproduce, grow, use energy, and respond to **stimuli** in the environment. All of these characteristics must be present and working together. Without one, all of the others would fail. Without a cell, there would be nothing to hold the DNA. Without DNA, the cell would not be able to use energy. Without energy, the cell would not respond to danger. If it were unable to respond, it would not live long enough to reproduce.

You are alive because your body is made of living cells, but you also depend on cells because you have to eat them to survive. Most food is alive. Crisp, juicy apples and fat, juicy steaks are alive. If they were not, they would be brown, rotten, smelly, and inedible. Cells have the ability to store energy so food stays fresh for a while, especially under refrigeration. Refrigeration slows down the life processes of the cells.

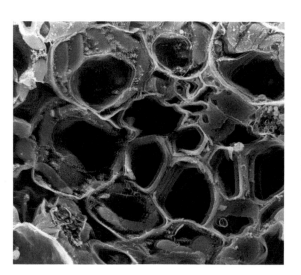

Cells in a leaf of maize, or Indian corn.

### DOWNLOAD

- Humans have 46 chromosomes, or 23 pairs (23 from the mother, 23 from the father).
- The twenty-third pair of chromosomes in humans determines their sex.
- Asexual reproduction produces offspring that are identical to the parent.
- Sexual reproduction produces a new individual, which differs from each parent.
- A living thing has cells, DNA, uses energy, responds to stimuli, grows, and reproduces.

## Viruses

The current definition of life is being challenged as scientists consider **viruses**, which contain a single strand of DNA, but are not made of cells. They need energy to grow, but they do not eat or make their own food.

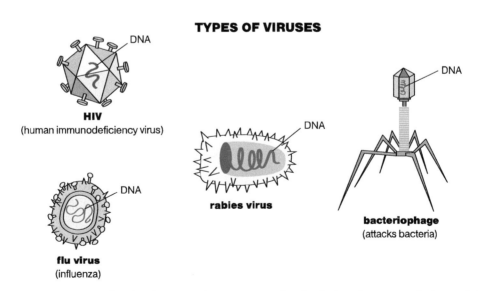

**TYPES OF VIRUSES**

Viruses are many times smaller than bacteria. They are not made of cells, but do carry small pieces of DNA.

Human immunodeficiency virus, or HIV, is a virus that attacks the immune cells of the human body. When the immune cells are destroyed, infected people cannot fight off diseases. Even common diseases, such as the cold, which people without the virus fight off easily, become life threatening for people with HIV. HIV causes the condition called AIDS, or acquired immunodeficiency syndrome. When a person carrying HIV begins to show symptoms of various diseases, they are said to have AIDS.

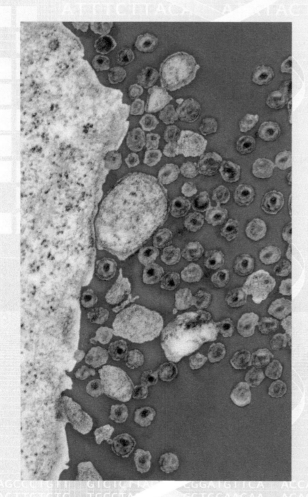

HIV viruses *(colored orange)* attack a human cell *(colored blue)*.

They can reproduce, but only if they have a living cell to do it for them. Like parasites, they invade living cells, inserting their own DNA into the DNA of the cell. The virus DNA tells the cell to make more viruses. Eventually, the cell bursts, releasing more viruses into the environment, where they invade more cells. So perhaps all something needs to be alive is to have working DNA.

## Bacteria

Bacteria are the smallest living things. Some are more than a thousand times smaller than the width of a paper clip. You might find tens of thousands of them in a single drop of water. Most range in size from 0.5 to 5 **micrometers** (1 micrometer = 1/1,000,000 meter or one-millionth of a meter) and require a microscope to be seen. Your own cells are about ten times larger than most bacterial cells.

Bacteria are all around us. They live in every environment possible: in dirt and water, in volcanic vents, in drain pipes, on pond scum. Some manage to survive in radioactive waste. They even live on and in your body! Bacteria help protect your skin and they help you digest your food. Over a thousand different species live in your intestines alone.

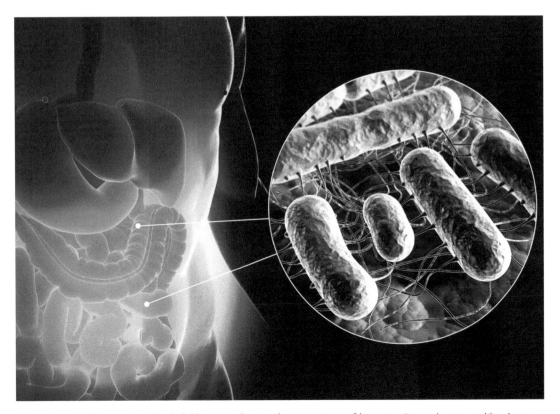

More than 1,000 species of helpful bacteria live in the intestines of humans. Some, however, like these salmonella bacteria, can cause food poisoning; others can cause disease.

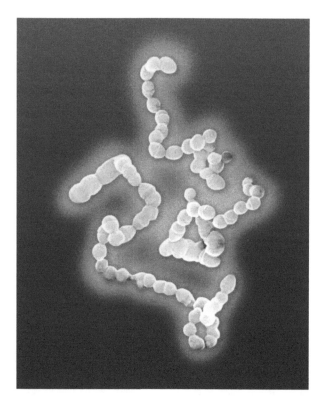

Streptococcus bacteria can cause strep throat, pneumonia, and even tooth decay. What shape are they?

Bacteria are also used to make foods, such as cheese, pickles, and yogurt. To make cheese, for example, producers add a special kind of bacteria to milk. The kind of bacteria they choose gives the cheese a distinctive flavor. The bacteria eat the milk sugars, **excreting** or giving off an acidic waste, which causes the milk to curdle into cheese.

**SHAPES OF BACTERIA**

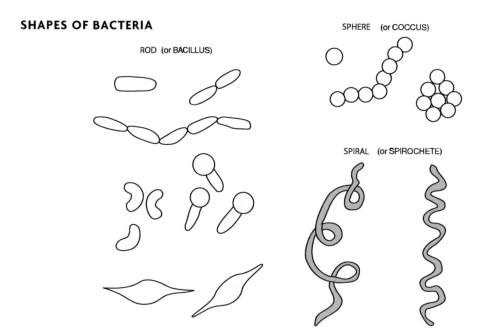

Most bacteria are grouped into three basic shapes: an elongated rod, a circular sphere, or a twisted spiral.

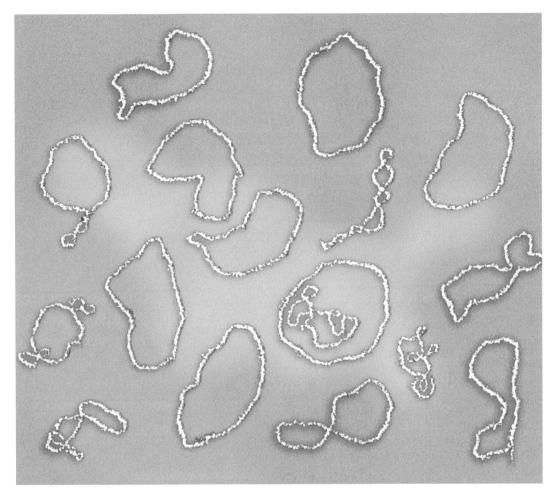

Plasmids, small extra rings of DNA found in bacteria (one per bacterium), look like rubber bands.

While most bacteria are harmless, even helpful, some do cause diseases, such as strep throat, pneumonia, and food poisoning. When bacteria cause diseases they are called **pathogens**, from the Greek words meaning "to cause suffering." Even the normal, healthy bacteria on your skin can be problematic if the skin is not kept clean. Bacteria thrive in warm, moist places, such as your armpits; the wastes they give off can build up and cause body odor.

There are about 5 nonillion types of bacteria. (That's a five with thirty zeros after it, or five trillion times a billion times a billion!) In your own body, they even outnumber your own cells ten to one. Bacteria come in every shape imaginable. Most look like beads or capsules. Others resemble springs, worms, kidney beans, or balloons.

Unlike other cells, bacteria have one circular strand of DNA—one chromosome—shaped like a rubber band. That chromosome can have up to 12 million base pairs. Humans have 46 chromosomes and about 3 billion base pairs. Besides their chromosome, bacteria also have small extra pieces of DNA called **plasmids**. Plasmids contain genes that help bacteria resist poisons and antibiotics. Scientists use plasmids in **recombinant DNA** technology to insert new genes into other organisms and study their effects.

## ASEXUAL VERSUS SEXUAL REPRODUCTION

Most plants and animals have more than one chromosome in each of their cells. They may reproduce asexually or sexually. **Asexual reproduction** requires only one parent. Males and females do not exist. A cell in the organism simply clones itself by making an exact copy of its chromosomes and passing them on to the next generation. If the cell starts with one chromosome, it will divide into two cells with one chromosome in each of the new cells. If the cell starts with two, each new cell will have two.

**CHROMOSOME NUMBER AND REPRODUCTION**

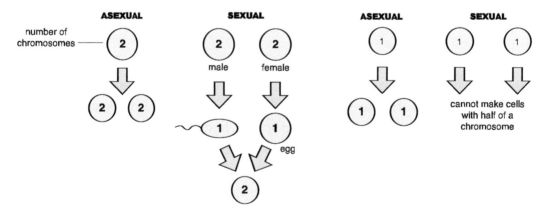

Organisms with an even number of chromosomes might reproduce asexually or sexually, depending on the species. Organisms with an odd number of chromosomes, however, must reproduce asexually, because their chromosomes cannot be divided exactly in half to make sperm or eggs.

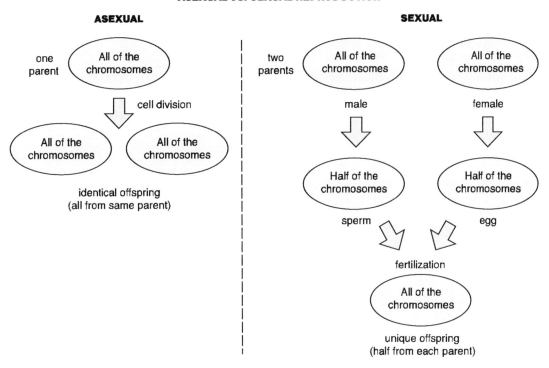

In asexual reproduction, a copy of the original adult is made. In sexual reproduction, half of the chromosomes from two different adults (a male and a female) mix to create a unique individual.

Any organism with an odd number of chromosomes has to reproduce asexually because in sexual reproduction the chromosomes have to pair up before the cells divide. If there were an odd number of chromosomes, one chromosome would not have a partner. When the partners separated to move into the new cells, one cell would end up with one chromosome less than it needs.

All bacteria and many plants and animals reproduce asexually. If an organism has an even number of chromosomes, it could reproduce either asexually or sexually. If the species had males and females, then each parent would pass on half of the chromosomes. If there were no males or females, then any individual would clone the even number of chromosomes that it had.

**Sexual reproduction**, therefore, requires two parents, a male and a female. The offspring they produce are not clones, because they get half of

**POP-UP**

Chromosome number varies from species to species. The following shows the number of chromosomes found in some commonly known organisms:

| | | | |
|---|---|---|---|
| Bacterium | 1 | Rabbit | 44 |
| Mosquito | 6 | Human | 46 |
| Kangaroo | 12 | Chimpanzee | 48 |
| Pea | 14 | Potato | 48 |
| Alligator | 32 | Skunk | 50 |
| Slug | 36 | Horse | 64 |
| Cat | 38 | Camel | 70 |
| Mouse | 40 | Chicken | 78 |

Chimpanzees have two more chromosomes than humans, but they share 98.4 percent of the same DNA.

## EXAMPLES OF REPRODUCTION

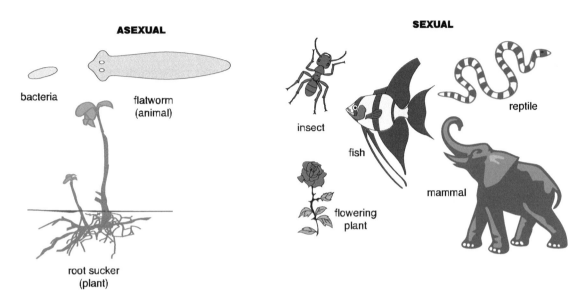

Bacteria and many plants and animals reproduce asexually. Flatworms divide in half to form two new flatworms, and many plants send out roots that sprout into new plants called suckers. Most complex organisms—such as flowering plants, insects, and higher animals—reproduce sexually.

their chromosomes from one parent and half from the other. The parents' DNA recombines to make a unique individual that is different, but also similar, because it shares characteristics from each parent.

Interestingly, the number of chromosomes in a species does not determine how **complex** it is. A potato has more chromosomes than you do, but it is obviously a simpler organism. A mouse has more than a cat, and a chicken has more than a horse or a camel. A **plankton** species called *Aulacantha* has only one cell, but has 1,600 chromosomes. Scientists believe complexity has more to do with the size of an organism and how many different kinds of jobs its cells perform than with the number of chromosomes it has.

## GENOMES

**A genome** is the complete set of genes that is found in an organism. Smaller organisms usually have smaller genomes, but just as with chromosome number, it varies. Humans have around 33,000 genes. A fruit fly has around 13,600. By comparing genomes, scientists have discovered that many of our genes are shared with other living things. For example, humans have about 40 percent of the same genes as a mouse, and more than 98 percent of the same genes as a chimpanzee. When scientists compare the genomes of different humans from around the world, they find that they differ by only 1 percent. That may not sound like much, but if you do the math, that means about 3,300 genes that make you different and unique from everyone else on Earth.

The U.S. government's Human Genome Project completed the cataloguing of all the bases on human DNA in 2003. The catalogue includes all of the genes on all of the chromosomes, including both of the **sex chromosomes**. The sex chromosomes, which are the twenty-third pair of chromosomes in humans, determine the sex of an individual. Male humans actually have about 900 fewer genes than females because one of their sex chromosomes, called the Y chromosome, is shorter. The genes on the Y chromosome are what give a male his characteristics. Females do not have a Y chromosome, so they have different traits.

The Human Genome Project showed that humans had about 33,000 genes. That surprised scientists because they previously believed humans

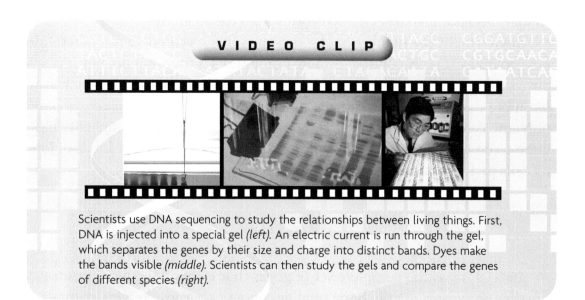

**VIDEO CLIP**

Scientists use DNA sequencing to study the relationships between living things. First, DNA is injected into a special gel *(left)*. An electric current is run through the gel, which separates the genes by their size and charge into distinct bands. Dyes make the bands visible *(middle)*. Scientists can then study the gels and compare the genes of different species *(right)*.

had to have at least 100,000 genes. They also discovered that only about 1.5 percent of our entire DNA actually regulates and makes proteins. The function of the rest remains to be determined. One of the continuing goals of the Human Genome Project includes using gene knowledge to study human biology and medicine.

Scientists use genomes to study the relationships between living things. Presumably, the more genes two organisms share, the more closely related they are. By studying the mutations and changes in the genes over time, scientists can trace the ancestries. With advances in technology, genome sequencing is getting cheaper and faster. Someday you may be able to find out how closely related you are to any other person on Earth simply by comparing your genome to theirs.

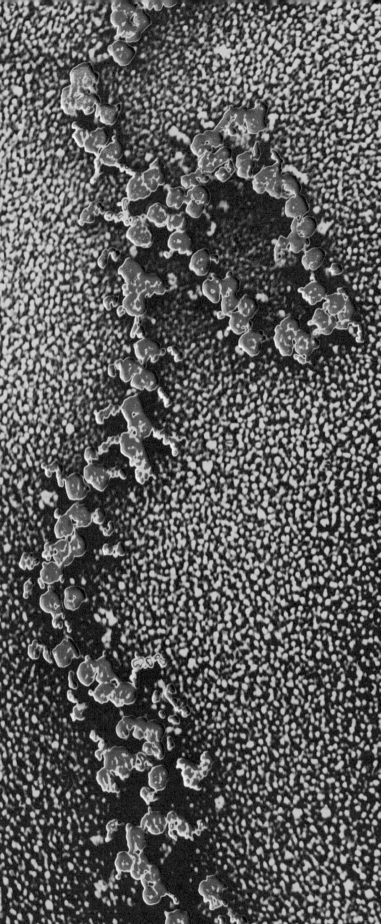

Ribosomes from an insect cell begin the process of protein production.

# CHAPTER 4

# The Language of the Genes

**You are** a spy, working undercover in a foreign land. Your job: to contact secret sources, gather intelligence, and smuggle it by code out of the country. Back at headquarters, special decoders receive the message and convert the code into ordinary language. Headquarters takes action, sending out supplies for the mission. Undercover work like this takes place inside cells, without rest.

## GENE SPEAK

**The process** of coding and decoding takes place between DNA and ribonucleic acid, or RNA. DNA is the source of the information. It carries the intelligence needed by the cell in the form of its base messages. A special kind of RNA called messenger RNA, or mRNA, is like the "spy." It converts the DNA message (reading only one side of the double helix at a time) into code by a process called **transcription**. Following Chargaff's rules, the mRNA code is complementary to the DNA bases. In genetics, complementary refers to the chemical attraction that one base has for

## TRANSCRIPTION, TRANSLATION, AND PRODUCTION

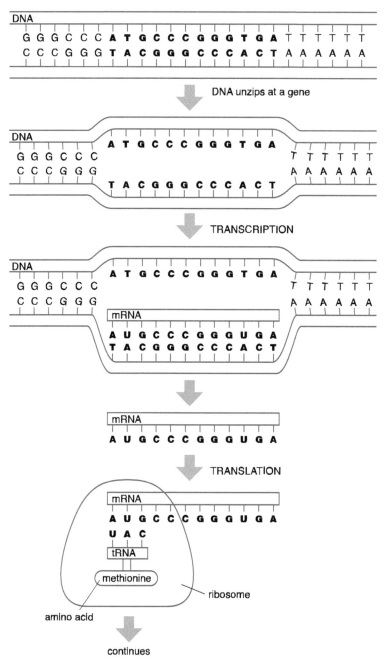

The production of a protein. During transcription, the DNA code is changed into an mRNA code. Once transcribed, the mRNA moves out of the nucleus to a ribosome in the cytoplasm of the cell. During translation, each triplet on the mRNA is decoded by a tRNA and a specific amino acid is added. During production, the process of adding amino acids in a string continues, producing a long protein.

## TRANSCRIPTION, TRANSLATION, AND PRODUCTION continued

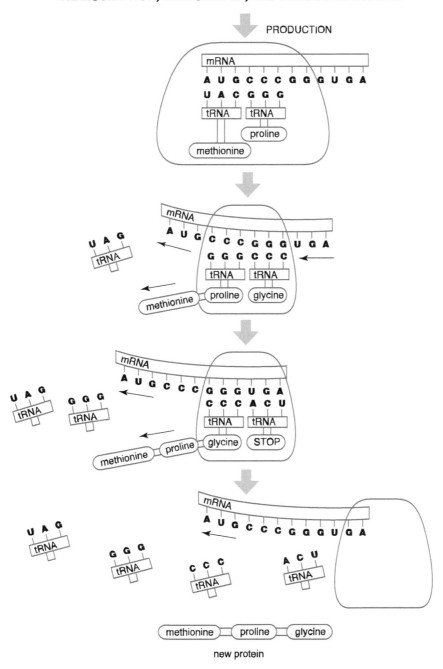

PRODUCTION

new protein

A stop codon signals the end of a protein and production stops. The new protein moves to where it is needed in the cell. A specific protein is made each time depending on the original DNA message.

another base. For example, the DNA triplet GGG would be translated into the mRNA codon CCC because each G would attract a C when the DNA unzips.

Once coded, the mRNA carries the message out of the nucleus to headquarters, the **ribosome**, where decoding takes place. Ribosomes are special organelles made of another kind of RNA called ribosomal RNA, or rRNA. The ribosome is the construction site for building proteins, which are the supplies needed for the mission of life.

When the mRNA reaches rRNA headquarters with its coded message, it binds with the ribosome and the decoders take over in a process called **translation**. The decoders are made of yet another kind of RNA called transfer RNA, or tRNA. Each tRNA holds a triplet of bases called an **anticodon**. Each anticodon is complementary to one mRNA codon, which means they carry the same message as the original DNA triplet. Attached to the other end of each tRNA is a specific amino acid.

The mRNA feeds its codons, three base letters at a time, into the ribosome. The first anticodon that matches with the first codon begins the protein chain with its particular amino acid. Then, the second anticodon matches with the next codon, attaching its amino acid to the first one. This continues as each codon is translated into a specific amino acid. For example, GGG on the DNA converts to the codon CCC on the mRNA. The anticodon GGG on the tRNA (which was the original message)

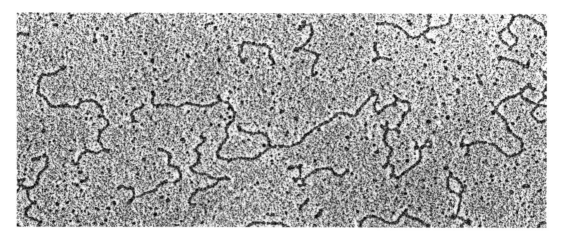

Ribosomal RNA (rRNA) in a human cell. Ribosomes are made of protein and rRNA.

matches with the mRNA CCC and the tRNA attaches an amino acid called proline to a growing protein.

Once an amino acid is attached, it releases the tRNA, which returns to the cell to find a replacement for the amino acid it lost. As the process continues, a string of amino acids is pushed out of the ribosome like toothpaste squeezing out of a tube. Production ends when a so-called stop codon causes a completed protein to break away from the ribosome. If a different message comes from the DNA, then amino acids are added in a different order, thereby making a different protein.

RNA and DNA are perfectly matched for cell messaging, but there are several important differences between them. Because RNA reads only one side of the unzipped DNA molecule, it is usually single-stranded. It has four bases like DNA, but not the same ones. They both have adenine, cytosine, and guanine, but instead of thymine, RNA has uracil.

Uracil is symbolized by U and is the complement of A. (For transcription, just picture DNA as a coach yelling, "'Ay, you! Get over here." In other words, A bonds with U. The triplet GAA on the DNA would transcribe into the codon CUU on the mRNA. Then CUU would be decoded by the anticodon GAA on the tRNA. The codon CUU tells the tRNA to attach leucine to a protein.)

**D O W N L O A D**

■ Proteins are long strings of amino acids.

■ Making a protein occurs in three stages: transcription, translation, and production.

■ Ribosomes are the centers of protein production.

■ Each codon on a messenger RNA signals the addition of a particular amino acid to a protein.

■ Proteins are used for nearly all cell activities.

The length of a protein is determined by the start and stop codons. The most common start codon is AUG. It always codes for the amino acid methionine. The codon UGA is a common stop codon, although others are used. For example, the mRNA code AUGCCCGGGCCCUGA would be translated one codon at a time, starting with the A, so it contains five codons. It is read as: AUG = start making a protein with methionine, CCC = add a proline, GGG = add a glycine, CCC = add a proline, and UGA = stop making the protein.

## BUILDING BLOCKS AND WORKHORSES

Like boards that make the frame of a home, proteins build cells. The more wood, the larger the home, and the more proteins the larger the cell. When one home is finished, builders move on to another. When one cell is built, it reproduces or divides and makes more cells. The human body grows by adding more and more cells. Proteins build muscles, causing them to get larger and bulge. Regardless of how much exercise an athlete gets, without protein he would not get any stronger.

The four different bases in three possible positions in a DNA triplet, or word, combine to form a total of 64 unique words. Those 64 triplet words can then be combined to make tens of thousands of different sentences or messages. This is what makes the DNA language so complex. Human DNA makes more than 100,000 different kinds of proteins.

You would think that 64 codons would code for 64 different amino acids, but proteins are actually made with combinations of twenty different amino acids. That is because some codons signal for the same amino acid. For example, AAA and AAG both code for the amino acid lysine. Six different codons code for the amino acid serine. Some proteins are small, built from only a few amino acids. Others are large, made from a string of more than 10,000.

For protein synthesis to occur, your cells must have a ready supply of amino acids. Your DNA can make ten of them. The rest, called **essential amino acids**, are supplied by your diet. This is why it is important to eat protein every day. Proteins are found in meat, fish, milk, cheese, eggs,

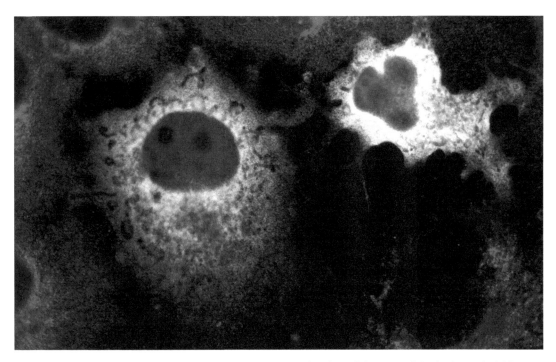

Special photography reveals proteins being produced inside of a cell *(upper right)*. The large dark blue spot is the nucleus. One protein, which appears green, is evenly distributed throughout the cytoplasm; another appears as clumps of light blue spots.

beans, nuts, and whole grains. The proteins you eat are then broken down, back into amino acids. Then your tRNA can pick them up and use them to make more proteins for your body. Without proteins, your heart, muscles, and other organs would weaken and you would succumb to disease.

Different amino acids are essential for different species. What people require in their diet differs from the requirements of a bacterium, plant, cat, or dog. In fact, the amino acid needs of cats and dogs are different, too. This is why cats should never eat dog food. It lacks the amino acids cats need to stay healthy. Pet food companies invest a lot of time and money developing special diets that include the amino acids needed for each pet.

Bacteria, along with many other single-celled organisms, take in nutrients through their outer covering, the cell membrane. In any case, their DNA makes all of the amino acids that they need.

Proteins are also the workhorses of cells. They keep things moving. Embedded in the cell membrane are little antennae-like structures called

**PROTEINS IN THE CELL MEMBRANE**

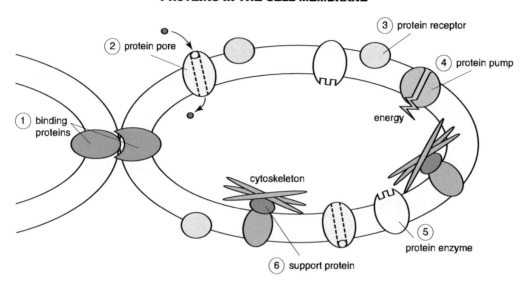

1. Binding proteins hold cells together; 2. Protein pores allow substances to move in and out of a cell; 3. Protein receptors receive chemical messages from the outside; 4. Proteins pumps use energy to pump substances across the membrane; 5. Protein enzymes help run chemical reactions; 6. Support proteins work with the cytoskeleton (cell skeleton) to give a cell its shape.

receptors, which are made of proteins. The protein receptors detect chemicals and food. Other proteins work in the cell membrane like trap doors, allowing what the cell needs to pass through and shutting out what might be harmful. Still others form root-like structures that grow into other cells and hold groups of cells together. Without these protein grips, you would fall apart and become a pile of single cells on the floor.

Proteins form **keratin**, the hard material that forms fingernails and toenails, as well as the hooves of a horse and the horns of a goat. Proteins can also be soft, as in your hair, the wool of sheep, and the feathers of birds. Proteins can be strong and elastic, used to form the **tendons** and **ligaments** that attach your muscles to bones and your bones together. Proteins make **hemoglobin**, which carries oxygen on your red blood cells, and enzymes like pepsin that speed up digestion of food in your stomach. Have you ever noticed that when you vomit, the back of your throat burns? That is the pepsin (along with stomach acids) that has come up from your stomach. It burns because it is digesting the cells in your throat.

ALERT !

Proteins may cause allergies! An allergy occurs when the immune system becomes overly sensitive to a foreign substance, something that would not normally be found in your body. The substances that cause allergic reactions are called allergens. Allergens are often proteins from another organism, such as a mold *(center)*, dust mites *(right)*, plant pollens *(left)*, animal hairs, or certain foods. Scientists are now making vaccines from some of these proteins. The vaccines can be used to slowly sensitize a person to a particular allergen so that they no longer react when exposed to it.

Proteins are an important part of DNA research. By studying different types of genes, scientists learn how each protein affects the way the body functions and grows. Scientists want to understand how genes determine which particular protein needs to be made. They want to know how to turn genes on and off artificially so they can make a beneficial gene work on demand or a defective (disease-causing) gene stop working. They want to know which gene processes are shared by different species. By speaking the language of the genes, researchers hope to capture the essence of life.

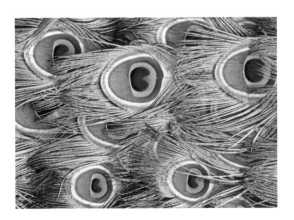

Feathers, like these of a peacock, are made of a special protein called keratin, which is produced by cells in a bird's skin.

Abnormalities in DNA can cause deformities in organisms like this sneezeweed flower.

# CHAPTER 5

# DNA Is Change

Imagine you are a shape shifter in a magical kingdom, changing at will into a charging horse or a soaring falcon. DNA behaves this way, allowing living organisms to change. You are a good example. You are not your mother; you are not your father. You are half of each, a new mixture of DNA that has never existed before. DNA can also change, or mutate, altering the characteristics of an existing individual.

## ATTACK OF THE MUTANTS

When you think of a **mutant**, do you think of scary, deformed monsters that go berserk and terrorize a city? Actually, mutations happen all the time. The DNA in your body has probably mutated several times just while you were reading this! Scientists estimate that as many as a thousand mutations may occur every time a cell copies its DNA. Most mutations are harmless, and your cells correct many that are not.

To understand how your body corrects mutations, you have to first understand how DNA **replicates**, or copies, itself. Most mutations occur dur-

While it might seem that this five-legged lamb was caused by a gene mutation, it is more likely that the extra leg came from an undeveloped conjoined twin (an identical twin that never separated from the twin that survived).

ing the copying process. DNA copies all of its bases during replication, when a cell divides or reproduces, but only part of its bases during transcription, when proteins are made.

When a cell gets ready to reproduce, special enzymes break the bonds that hold the bases together in the middle of the double helix. The double helix unzips, which separates the two sides and exposes the bases. New complementary bases (attached to their nucleotides) move in from supplies residing in the cell and bond with the exposed bases. An exposed A would attract a nucleotide carrying a T, and so on.

Imagine a dance with a line of people connected at the waist by a rope. They are facing another line of people connected the same way. Each line represents the phosphate-sugar sides of the double helix. Each person is holding hands with the person standing opposite. The arms of the people

**D O W N L O A D**

- A mutation is a change in one or more of the bases on a DNA strand.
- Mutations can be harmless, harmful, or beneficial.
- A mutation can allow a species to become better adapted to its environment.
- The three basic types of mutations are substitutions, insertions, and deletions.
- Mutations can be passed on through inheritance.

represent the bases. The linked hands represent the hydrogen bonds between the bases that hold the two sides together.

If one person after another dropped their partner's hands, the two lines would separate just like a DNA molecule when it unzips. Once unzipped, the people would no longer have partners. They are like the exposed bases. If a new group of people moved in, lined up, and held hands with the people on one exposed side, they would form a new double helix. If a new group of people moved in and held hands with the people on the other exposed side, they would form a second double helix. One strand of DNA becomes two.

In the case of DNA, however, each new strand is an exact copy of the original strand, assuming there were no mutations. For example, if A and T are paired along a strand of DNA and they are unzipped, A is exposed on one side and T on the other. T bonds with the exposed A, and A bonds with the exposed T. In this way, the strand starts with one pair, A-T, and ends up with two, A-T and A-T. This happens hundreds of thousands of times along a DNA strand, depending on its length, until there are two identical strands. When the cell containing the replicated DNA divides, each new cell gets its own perfect set of the original DNA.

When proteins are made during transcription, the DNA code from one gene is transferred to a strand of mRNA. The DNA only unzips at that one spot to make a particular protein, although the unzipping process (during

## DNA REPLICATION

DNA replication forms two DNA strands that are exactly like the original. Each new strand is made of half the original strand and half the new strand.

both transcription and replication) can occur at many different places along the DNA at the same time. This speeds things up so the cell can reproduce faster and perform more jobs than if it copied only one section at a time. Without this ability, your body would slow down. If you got sick, you might not be able to make white blood cells fast enough to fight disease.

In both replication and transcription, the double helix is opened to expose the bases. This exposure is what allows mutations to occur. Mutations can alter the order of the bases on DNA or RNA. A change in the order changes the message. If you were told to feed the cat, you would know what to do. But if the sentence mutated and just one letter was changed, you might not do the right thing. You could be told to feed the fat. Then you might eat too much and the cat would starve!

## MUTATIONS: GOOD OR BAD?

A mutated DNA message can cause the cell to do the wrong thing. Cells depend on thousands of proteins to function properly. If a mutation alters a protein that is essential to the cell, a **genetic disorder** can occur. A genetic disorder can change the way an individual looks, as in albinism, or the way the **tissues** and **organs** work, as in cystic fibrosis. Albinism is a lack of pigment in the skin, which causes the skin to look white. Cystic fibrosis causes the lungs to produce a thick mucus, making it difficult for the affected person to breathe. Some genetic disorders even cause death. Disorders may be **inherited** from an ancestor, or caused by something in the environment, such as exposure to lead or pollution.

If a mutation occurs in the body cell of an organism that reproduces sexually, it cannot be passed on. Heart cells, stomach cells, and skin cells are examples of body cells. But if a mutation occurs in a sperm or an egg, then the new individual that is formed could inherit it. In species that reproduce asexually, such as bacteria or many plants, mutations are always passed on because a body cell is cloned to make the new individual.

After the completion of the Human Genome Project, researchers were surprised to find that long stretches of DNA seemed to have no function. Only about 5 percent of your DNA codes for proteins. Another small per-

People with the most common form of albinism have very light skin, hair, and eyes because a single gene change does not allow their bodies to produce normal amounts of a color pigment called melanin.

centage helps regulate, or control, your genes, telling them when to be active and make a protein or when not to work. The greatest percentage, sometimes called junk DNA, does not seem to have a purpose. Of course, it may not be junk at all. It may be that scientists simply have not yet discovered what its job is.

Some scientists have suggested that extra DNA and its ability to mutate actually helps a species to **adapt** to its environment. First, a mutation would have to occur in the junk DNA, leading to a new working gene. The new gene would produce a new protein and thus a new trait. Then, if the environment changed and the new trait made it more likely for the mutated individual to survive, that trait would be passed on to its offspring. Then the offspring would be more likely to survive and pass the trait on to their offspring. Over time, more and more individuals in the population would carry the new gene. Those without the mutation would die off. In this way, some of the population would be lost, but the species itself would not become extinct. It would instead be adapted to its new environment.

The rate at which a species mutates varies from species to species. The faster the DNA mutates, the faster a species adapts to its environment. Bacteria mutate very quickly, and are therefore able to develop immunity to antibiotics. If a single mutation in just one bacterium resists a particular antibiotic, the rest of the bacteria may be killed, but that one will not. That single bacterium will survive, reproduce, and pass on the mutation. Eventually a new population of bacteria will exist that cannot be killed with that antibiotic. Scientists then have to develop new antibiotics.

To prevent bacteria from becoming resistant in your body, you should not take an antibiotic unless it is absolutely necessary. When you do have to take one, take every last dose until it is gone. Some people quit taking antibiotics when they feel better, but this only kills some of the bacteria and leaves those with resistant genes to adapt and become immune. If the person relapses or gets sick again the antibiotic will no longer be effective.

## THE MANY FACES OF MUTATIONS

**Mutations** occur when mistakes are made during replication or transcription, when bases are exposed to the environment of the cell. A number of things can occur: The wrong base may be copied, one or more bases

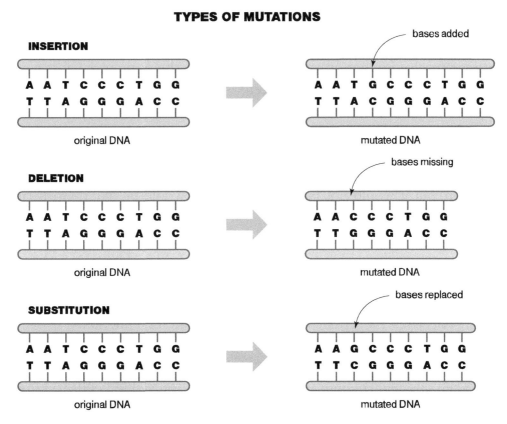

The basic types of mutations. All three types change the original message into a different message.

may be clipped out, or extra bases may be added where they do not belong. Entire genes may be copied more times than they should, or one or more genes may be eliminated. Mutations can occur in one or many places at the same time. They can even occur in mitochondrial DNA.

### Substitution Mutations

Mutations can be roughly **classified** as three main types: substitutions, insertions, and deletions. In a substitution, one base is exchanged for a different one. In the English language, feeding your fat instead of your cat would be an example: an f was substituted for a c. In the DNA language, exchanges between A and G, and between C and T, are the most common. The change in the base sequence would cause it to code for a different amino acid.

Not all substitutions cause a change in an amino acid. Many codons are **redundant**, meaning that two or more codons code for the same amino acid. For example, the codon AGU codes for an amino acid called serine. But if a substitution causes AGU to mutate to AGC, AGC also codes for serine. During protein production, serine is still put in the same place as it would have with AGU, and the same protein is made. Scientists refer to these as silent mutations, because they have no effect on the individual that carries them. It would be like baking cookies with dark chocolate instead of milk chocolate. Either way, you still end up with chocolate chip cookies.

If a substitution does change an amino acid, the message might still make a protein, but it would not be the right one. It would be like pouring orange juice on your cereal instead of milk. You would still have your breakfast, but it would be weird. In humans, a blood disease occurs when a substitution mutation causes the codon GAA to change to GUA. The tRNA then substitutes the amino acid valine for glutamic acid. The resulting protein causes normal red blood cells to collapse into the shape of a sickle or crescent moon. This causes a genetic disorder called sickle-cell anemia.

Normal red blood cells use a special blood protein to carry oxygen throughout your body. The oxygen is used to make energy. In a person with sickle-cell anemia, the mutated protein causes the sickled cells to burst and plug blood vessels. Oxygen is not able to reach the cells to make energy, so people with sickle-cell anemia have trouble exercising. They tire easily,

**POP·UP**

Sickle-cell anemia is a genetic disorder that causes normal red blood cells *(right)* to collapse into an odd, sickle shape *(left)*. A sickle is a flat, curved blade that was once used to cut crops.

gasp for breath, and experience severe pain—all because just one of 287 amino acids in a blood protein has mutated.

A substitution might also change a stop codon, causing translation to stop too soon or fail to stop. The protein ends up either too short or too long. Have you ever heard someone stop halfway through a sentence, or listened to someone who babbles on about nothing in particular? They make no sense. Stop codon mutations—whether substitutions, insertions, or deletions—are called nonsense mutations.

As with any mutation, a substitution can help a species become better adapted, giving it a survival advantage over other species. Recently, scientists discovered a human gene mutation that gives humans the ability to speak, while our closest relative, the chimpanzee, cannot. They compared the same gene in humans, mice, monkeys, and apes. They found that humans have a double substitution on the gene—a place where two single bases have changed—while the other animals do not. Scientists know it is a speech gene because people with another mutation of the original double substitution mutation have severe language problems. Even mutations can mutate!

One of the scientists who discovered a human speech gene talks about how mutations in the gene gave people a greater ability to communicate verbally:

> This is hopefully the first of many language genes to be discovered. . . . It is compatible with the hypothesis that language could have been the decisive event that made human culture possible.
>
> —Wolfgang Enard, BBC News Online, 2002

## Insertion and Deletion Mutations

In an insertion, a new base that was not there before is added to the DNA string. This causes a change in what is called the reading frame. The reading frame determines how the triplets on the DNA or the codons on the RNA are read. In the DNA sequence CCCTTTCCC, the reading frame would start with the first C and read three bases at a time. The three triplets would be CCC, TTT, and CCC.

But if a G is inserted after the first C to form CGCCTTTCCC, the reading frame is shifted. It would be like someone cutting in front of you in the cafeteria line. You and the rest of the line would get pushed back. Reading from the first C, the triplets would be shifted to CGC, CTT, TCC, and C. Three totally different amino acids would be attached together and the last C would not add an amino acid at all. A different protein would be made. Instead of feeding the cat, you might fleed ther cat. Whatever that is, the cat probably wouldn't like it.

Even more drastic is a "stutter" mutation in which the same three bases are inserted into a gene over and over again. You would be told to feed feed feed feed feed feed the cat. Enough already! The same thing happens in Huntington's disease. A repeat of the codon CAG tells the tRNA to add too many glutamic acids to its protein. The resulting protein kills brain cells. Eventually, people with the disease lose their ability to move and think normally.

In deletion mutations, one or more bases are removed from the DNA. As with insertions, the reading frame is shifted and the message makes no sense. Now you might eed the at. The poor cat is gone and you're left with an "at."

## REVERSING MUTATIONS

**Many mutations** are reversible. In humans, about fifty different enzymes act like building inspectors, constantly patrolling DNA and RNA to check for errors. When an enzyme locates a substitution error, it chemically breaks the bond holding the base error in place and replaces it with the correct base. An enzyme can snip out the extra base in insertion mutations, but deletion mutations are permanent and cannot be repaired by the cell.

Even if a mutation is not corrected and the wrong protein is made, the cell has some ability to fix the problem. Other enzymes can break down the bad protein into harmless, smaller pieces that can then be recycled by the cell. Other enzymes have the ability to destroy an entire cell so the mutation dies with it. By correcting mutations, your body protects you from disease.

Small mutations, such as those involving just one base, are easier for your body to correct. Mutations that affect an entire gene, groups of genes, or entire chromosomes are much harder to fix. Sometimes when the DNA replicates, an entire gene is copied more than once. The power of the gene might then increase. This can be either good or bad. For example, a protein that normally helps the heart might damage the heart if the cell makes an excess amount of that protein. On the other hand, if an increased protein helped the heart to work more efficiently, greater strength and endurance would result.

If entire genes are deleted, certain proteins can no longer be made. Sometimes entire genes change positions on the DNA, fusing with new genes and creating new proteins. Whole sections of DNA may be cut out,

Smoking is not only hazardous to a smoker's health, but to anyone around who inhales the fumes from the burning cigarette.

bringing together and fusing two genes that were once far apart. One particular fusion mutation in humans makes a protein that causes a kind of brain cancer.

Scientists have identified thousands of genetic disorders. Some are naturally occurring and others are caused by the environment. Ultraviolet light from the sun, x-rays used in medicine, gamma rays given off by radioactive materials, poisons used to control pests, and pollution in the air and water are all **mutagens**. A mutagen is anything that causes a mutation.

Cigarette smoke is also a mutagen. It contains 69 known chemicals that can cause **cancer**. Its mutagens include formaldehyde, which is used to embalm dead bodies; arsenic, a poison used to preserve wood; polonium-210, a highly radioactive material; chromium, which is used to make paint; benzene, which is used to make gasoline; and 1,3-butadiene, used to make rubber. If someone handed you a nice, thick blended shake of paint, gasoline, formaldehyde, and wood preservative with a few chunks of rubber floating in it, would you drink it? Why, then, would anyone smoke a cigarette?

Scientists are now creating their own mutations. By adding or removing one gene at a time in an organism, such as a bacterium, they can observe how the mutated organism changes. Bacteria make good subjects because they are small, plentiful, and a lot is known about them. Would you be willing to change your genes if it would cure a disease? Would you change them to become taller, smarter, or stronger? Whether you agree with it or not, scientists predict that genetic medicine will be one of the fastest growing industries in the future.

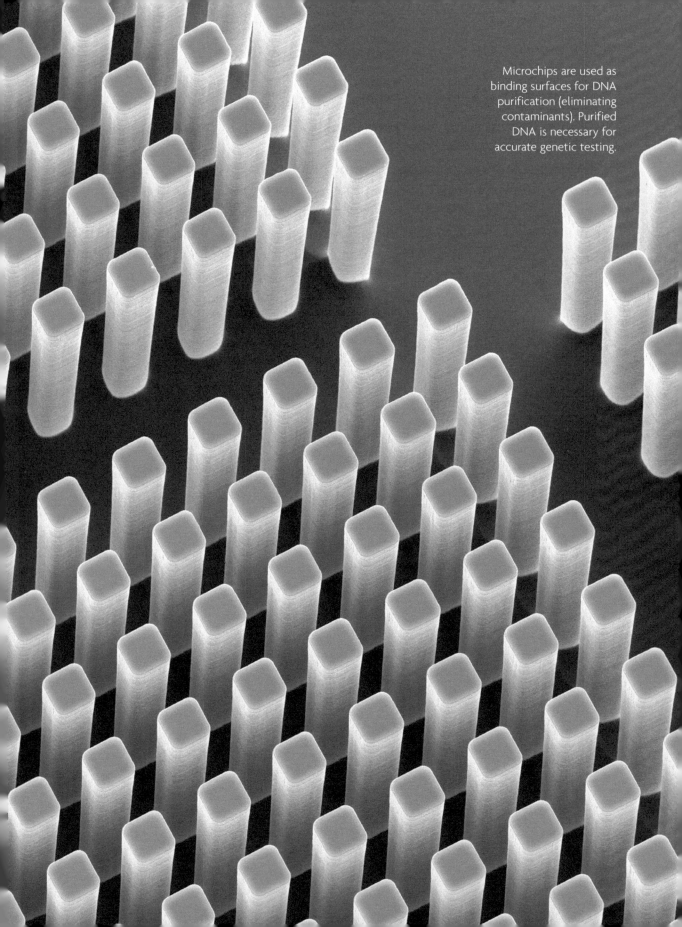

Microchips are used as binding surfaces for DNA purification (eliminating contaminants). Purified DNA is necessary for accurate genetic testing.

# CHAPTER 6

# DNA Now and When

**Currently,** DNA research focuses on two main areas of science: pure science and applied science. Pure science is the use of the scientific method. Scientists ask questions, do experiments, and find answers. Their answers make them think of more questions, which they ask to come up with more answers. The process is repeated over and over. These scientists may not directly do anything with their answers, but their research paves the way for applied scientists.

Applied scientists try to figure out how pure science can be used to help people, other organisms, or the environment. Over the years, they have found ways to fight disease, clean up the environment, solve crimes, identify relationships between people and other living things, save endangered species, and study fossils.

## TAMPERING WITH NATURE

**Many of** the foods you eat have had their genes modified, or changed slightly, through **genetic engineering**. Genetic engineers take a gene out of one organism and put it into the cells of a different organism. Millions of acres of corn in the United States have had their genes altered.

The new corn gene came from a bacterium, which makes a protein that is poisonous to insects, but not harmful to humans. Once the bacterium gene for that protein is injected into corn, it becomes a permanent part of the corn's DNA. When the corn multiplies, it passes on the new gene to all of the new corn cells. The corn then makes the poison. When an insect tries to eat the corn, it dies. Farmers no longer have to spray their crops with poisons that could contaminate the environment and hurt people and other animals.

The first genetically engineered food was the tomato. It received a gene that slows down the ripening process, so it stays firmer longer and will not rot before it gets to market. Other modified foods include strawberries, bananas, sweet peppers, squash, potatoes, and rice. Scientists are also modifying fruits and vegetables to make medicines or extra vitamins.

The meat you eat may soon come from cloned cattle that make leaner, juicier burgers and steaks. To make a cloned calf, scientists take the genes out of the body cell of a desirable cow, steer, or bull— one that has traits they want. Then they put the genes into a cow's egg that has had its DNA taken out and put the cloned cell into the uterus of another cow as if it were a fertilized egg. A cloned calf is born that is identical to the original cow

A calf is born by normal reproduction to a cloned cow.

or steer. Herds of cloned cattle are already being created in the United States. Because the cattle are identical, their meat is the same.

## DNA COMPUTERS

DNA computers are nothing new. You have had them working inside of you ever since you were created. In fact, they have been around since life began. Now, scientists have created DNA computers in test tubes that are 100,000 times faster than the fastest known electronic computer and run on their own fuel. Although they are not yet available to the public, they have the potential to change computing in the future.

Currently, computers use electrical signals (sent through silicon microchips) to create switches. The switches can be turned on or off to create a code. Different combinations of the code create the instructions to run the computer.

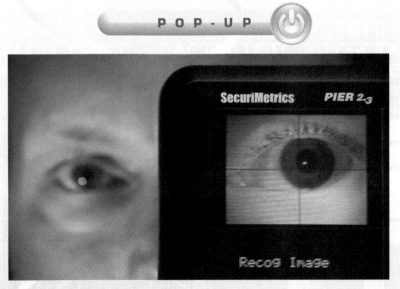

Like a contraption out of a James Bond movie, people may one day use DNA scans to verify their identity. No disguise can cover up the uniqueness of your genes. Scientists have already developed heat-sensitive thumbprint readers and iris scanners of the eye (above) as alternatives to keys, locks, or secret combinations, thus limiting entry into high-risk or top-secret areas.

DNA computers, however, are made of genes, which produce proteins that react chemically with one another. The chemical reactions can be used to create the on and off switches that code a computer program. To change the computer's program, scientists simply change the genes they use.

DNA computers are smaller and faster than any computer ever made, and they are environmentally friendly. To make microchips for traditional computers, the mineral silicon has to be mined and processed, which pollutes the environment. DNA can be synthesized, or produced, in a laboratory test tube by chemically stimulating it to make copies of itself. And because DNA computers can be programmed to power themselves, electricity is not required to run them. Perhaps one day, you will have a tiny DNA computer built into the side of your pencil. That would come in handy in math class.

## DNA DISEASE FIGHTERS

**Treating diseases** with DNA is called gene therapy. The possibilities for its use in medicine seem endless. Scientists are already experimenting

with genes that can be placed in cows so they produce medicines in their milk. They have also discovered that fruit flies, which are plentiful and easy to study, share about 70 percent of the same disease genes that humans have. Studying fruit fly genes could lead to a better understanding of human diseases, and might even help people live longer. A gene mutation discovered in some fruit flies, called the "I'm Not Dead Yet" gene, allows them to

The common fruit fly (*Drosophila melanogaster*) is often used to study genetics.

live twice as long as normal flies. Understanding that gene could lead scientists to gene therapy for extending human lives.

## Preventing Disease with DNA

Scientists are also working to develop DNA vaccines for such diseases as the bird flu, West Nile virus, and anthrax. The bird flu virus, which has devastated chicken and other bird populations, can spread to humans. The deadly West Nile virus infects both humans and horses. A DNA vaccine for horses has been approved, but one for humans is yet to come.

Anthrax is a contagious disease of cows and sheep that is caused by a bacterium. Infected animals develop sores on their body and in their lungs. They often die. Scientists are now testing a DNA vaccine against the disease by injecting pieces of the bacterium's DNA into mice. The anthrax DNA becomes part of the DNA of the mice. The mice then produce an anthrax protein, which triggers an immune response in their bodies. After this treatment, the mice do not get sick when exposed to anthrax. Their bodies have already built up immunity against the disease and they are able to fight it off.

Scientists believe that DNA vaccines will be much safer to use than traditional vaccines. Traditional vaccines use weakened or killed whole organisms, which can cause side effects. Rarely, they may even cause the disease that they were supposed to prevent. DNA vaccines contain genes that make proteins,

Scientists are using genetic engineering to produce vaccines against serious diseases.

which are more easily recognized by the immune system, allowing it to function more normally without side effects. Because of the reduced risk of side effects, more than one DNA vaccine can be given at a time. DNA vaccines are also cheaper to make and easier to store because they do not require refrigeration. Scientists are already working on DNA vaccines for humans against some viruses, a blood disease called malaria, a lung disease called tuberculosis, and certain types of tumors.

## Curing Disease with DNA

Tumors are cancers that need a lot of blood to grow and spread, so scientists at Scripps Research Institute in the United States decided to target a tumor's blood vessels. They engineered bacteria to make a protein found

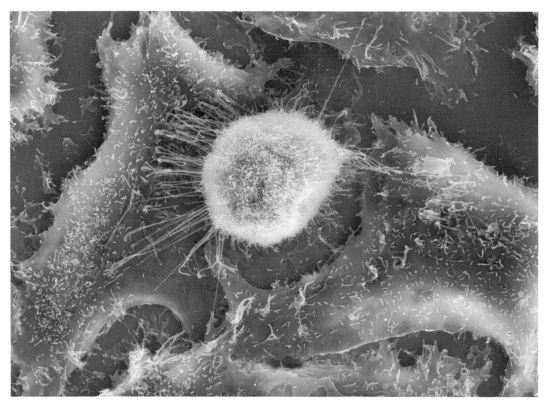

A human lung tumor (*colored yellow*) among healthy lung tissue. Genetically engineered bacteria may one day be commonly used to fight such tumors.

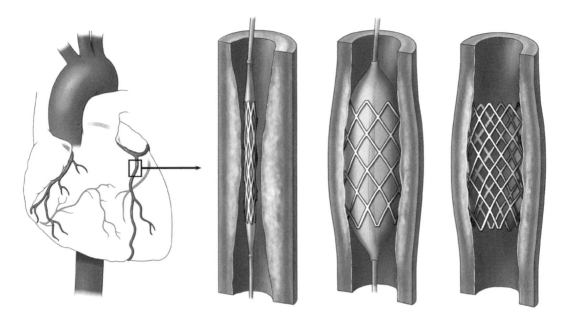

Current therapy for heart attack patients uses a stent (a cage-like structure) to open clogged arteries. In the future, gene therapy will encourage new arteries to grow around any blockages.

only in tumor blood vessels. When the engineered bacteria are put into mice, they trigger an immune response, but only against tumor blood vessels and not healthy vessels. White blood cells in the mice attacked the tumors, destroying their vessels and cutting off their blood supply. Without blood, the tumors died. If this technique can be perfected for humans, not only would it cure many cancers, but also patients with cancer would not have to endure traditional treatments like chemotherapy.

A similar therapy has been used successfully in people with heart disease. Thousands of genes that stimulate the growth of new blood vessels were grown in test tubes. The genes were then injected into the hearts of people with clogged arteries. New, healthy blood vessels grew, bringing oxygen to the diseased heart. Patients felt better, had more energy, and were even able to exercise again.

Gene therapy has been used to cure people with a gene mutation that causes them to regularly get fungal and bacterial infections. Researchers took white blood cells from the patients and injected them with healthy versions of the gene from a modified virus. Then they used chemotherapy to kill the defective white blood cells still in the patients' bodies. Finally, they injected the engineered cells back into the patients. Soon after, 20 percent of the patients' white blood cells carried the healthy gene, which was more than enough to cure their disease.

Genetic engineering may even one day help people with mental illness or depression. A promising study has shown that when just one gene from a social animal called a vole was injected into antisocial mice, the mice became more agreeable and loving. But while gene therapy holds great promise, there have been a few problems.

Scientists used the healthy gene of a virus to cure a disease called severe combined immunodeficiency disease, commonly known as Bubble Boy Disease. Children with the disease have a mutation in the gene that would normally give them an immune system. Without the healthy version of the gene, they easily catch diseases. Even a little cold, which most people can fight off, could be deadly for someone with the mutation. They have to live inside of a bubble—a sterile, plastic room that keeps them from being exposed to other people. While the therapy worked in many

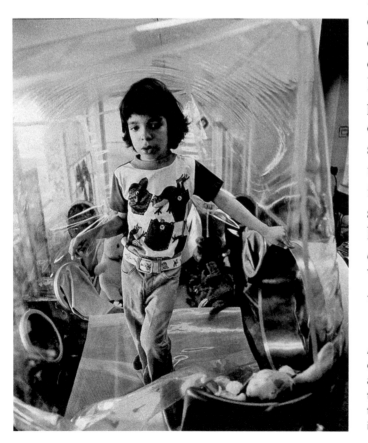

A genetic disorder commonly called "Bubble Boy Disease" leaves a person with no natural immunities against disease, confining them to a sterile environment inside of a plastic bubble.

patients, a few developed cancer. Scientists worried that the modified virus might have caused the cancer. Although no one had ever had that reaction to a gene therapy before, scientists decided it would be better for now to stop the experiment.

## DNA Testing

Now companies are developing DNA-based tests that can identify a mutation before a person gets sick. Early awareness of mutations can reduce the chances of illness, because knowledge of them allows people to take preventative steps, such as changing their diet or using gene therapy to correct the problem. Some DNA tests look directly at the bases on DNA. By comparing the tested DNA to a normal set of genes, they can spot mutations. Other tests look for the proteins or enzymes that are made by mutated genes.

DNA tests could also identify a person who is carrying a mutation and is unaware of it. Just as with Mendel's peas, people can have a hidden, recessive gene. They may not be sick (because they would also have one dominant form of the gene to keep them healthy), but they could pass the mutation on to their children. If the child happened to receive the same hidden recessive from the other parent, the child would have two recessives and they would get sick. They would not have a dominant form like the parents to keep them healthy. People who carry hidden recessives are called carriers.

DNA carrier screening can help parents who do not want to pass on harmful or deadly mutations. Once a defective gene is located, the parents' eggs and sperm are collected and the eggs are fertilized in the lab. Developing **embryos** are checked to see if they have **inherited** the defective gene. If not, a healthy embryo is placed back into the mother, who will then experience a normal pregnancy. The same method makes it possible for people to choose desirable traits to be passed on to their children. They might choose their child's height or eye color. Some people want to be able to choose their children's traits. Others are against it and believe that the creation of children should be left to nature.

Hundreds of DNA diagnostic tests are now available for people to use. They can determine if a person carries genes for Lou Gehrig's disease or muscular dystrophy, which cause muscle weakness and paralysis; cystic fibrosis, which results in coughing and lung infections; or the bleeding disorder hemophilia. The blood of a person with hemophilia does not clot, so they can easily bleed to death if cut. Other helpful tests are being developed every day.

## SAVING AND CREATING LIFE

Genetic engineering could be used in **bioremediation**, the process of using biological organisms to clean up the environment. Some organisms have the ability to break down or digest wastes in the water, air, or soil. If scientists can identify them, the organism's genes could be engineered to make them even more effective at those jobs. Extra genes could be added,

A contaminated lagoon at an industrial waste facility is being treated through bioremediation, which uses bacteria to digest toxic sludge.

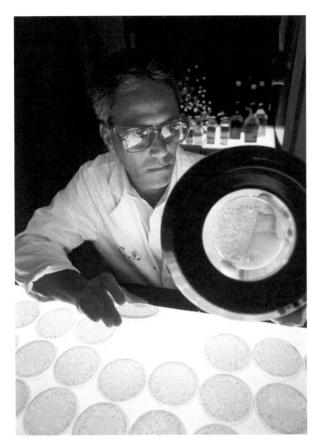

Pollution-eating bacteria are grown and studied in a laboratory.

so the organisms could break down more than one kind of toxin and do more than one job at a time. Bacteria could be engineered to break down human garbage, restoring the land that is now used for landfills. Scientists have already created waste-eating bacteria, but they have not won government approval to use them. The government wants to be sure that the altered bacteria will not create a problem in the environment if they are released.

In the meantime, scientists continue to create different kinds of waste-eating bacteria in the lab. They are now working on a modified bacterium that not only cleans pollution from the air, but also **excretes**, or gives off, a beneficial waste in the form of hydrogen. Hydrogen, which does not pollute when it is burned, can power machines and automobiles. Humans could preserve life on Earth by reducing global warming and their dependence on oil and gasoline at the same time!

DNA can also save endangered species, and possibly even bring back extinct species. It is generally believed that extinction is forever, but genetic engineering may change that. Scientists could place a complete set of DNA from an organism that is disappearing, or has disappeared from Earth, into the cell of a close living relative. The relative could then produce a clone of the endangered or extinct organism. Scientists are already attempting this with endangered species, such as the gaur, an ancient relative of the

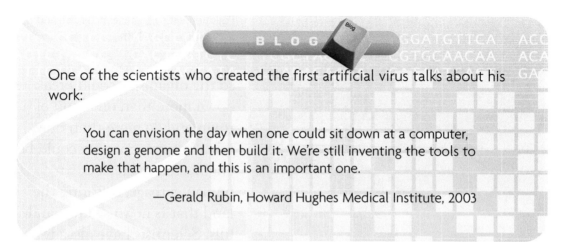

One of the scientists who created the first artificial virus talks about his work:

> You can envision the day when one could sit down at a computer, design a genome and then build it. We're still inventing the tools to make that happen, and this is an important one.

—Gerald Rubin, Howard Hughes Medical Institute, 2003

cow. A cloned gaur cell was placed in the uterus of a cow and the cow gave birth to a gaur. Although currently unsuccessful, scientists are also trying to bring back an extinct animal called the quagga, which looks like a cross between a zebra and a giraffe, and an extinct dog-like **marsupial** called the Tasmanian tiger.

Scientists are even creating life from scratch, making new organisms that have never existed before. They started by building a virus, using a known virus as their model. Their goal was to determine the least number of genes needed to support life. The genes were ordered from a company that makes artificial genes and fused together to make a chromosome.

Once the synthetic virus was built, it was injected into a bacterial cell, and the cell began to produce more viruses. It needed only 300 genes to make the virus function normally. Compare that to the 33,000 genes needed to sustain human life. But was new life really created? By current definition, viruses are not alive. Either way, scientists feel that this is the first step toward creating the first artificial bacterium.

## Who Done It?

DNA has been used for some time in the field of **forensics**, which uses science to solve crimes. When crimes are committed, investigators gather evidence in the form of poisons, weapons, dental records, footprints, blood type, fingerprints, and now DNA. They value DNA evidence because it

**TOOL   BAR**

Because DNA sequencing creates a DNA "fingerprint" that is unique to an individual, it can be used to identify a suspect or a victim of crime.

can be collected from a single hair or a tiny drop of blood. It has also been shown to be more reliable for identifying criminals than an eyewitness. DNA can also identify the victims of crimes, even when the only evidence is a single bone.

DNA evidence is analyzed at special labs. Lab workers called technicians study a dozen different areas on a DNA sample. They choose areas that would most likely be different from person to person, such as those with the genes for hair color or lip size. They do not choose areas that are the same in most people, such as those responsible for making a brain or a heart. From their **data**, they create a profile called a DNA fingerprint.

Once investigators have a DNA fingerprint, they can collect a DNA sample from a suspect, make another DNA fingerprint, and compare the two. The more gene sequences that match on the profiles, the stronger the evidence is against the suspect. DNA evidence is powerful because the likelihood of any two suspects having the same exact profile is extremely small.

The innocent are also protected by DNA fingerprints. If there are no matching sequences, then the suspect could not have left the evidence at the crime scene. With recent advances in DNA fingerprinting, a number of people have even been released from prison and cleared of crimes they did not commit. It is predicted that in the future investigators will be able to compare a person's entire genome, instead of the handful of sequences they use now.

DNA fingerprinting can also be used in medicine to identify organ donors for people who need transplants. By choosing a donor with the closest DNA match, the recipient is less likely to reject the donated organ. Even more amazingly, scientists hope to eventually be able to grow entirely new, healthy organs inside the bodies of patients who need transplants. By using a patient's own DNA, new organs could be grown to replace the diseased ones. Once the new organ was ready, a surgeon could remove the old one and transfer the new into its place. The patient's body would not reject the new organ and they would not have to take anti-rejection drugs for the rest of their life.

Medicines could also be tailored to fit individuals. A DNA fingerprint would tell doctors if someone was likely to have a bad reaction to a particular drug. If the test was positive, a different medication would be prescribed. DNA screening is already available for determining a person's potential reaction to several prescription and over-the-counter drugs.

Even historians are getting in on the act. They are using DNA to solve the mysteries of the past. In one case, a woman claimed to be Anastasia, the daughter of one of Russia's imperial families. The entire family had presumably been executed in 1918. DNA evidence conclusively proved that the woman could not possibly be Anastasia.

Wildlife officials are using DNA fingerprinting to prosecute poachers who hunt rare and endangered species illegally. Just a few cells from a bit of skin or piece of meat are all they need to identify an endangered species and make a case.

Some city officials have begun using DNA fingerprinting to identify polluters. If sewage is not disposed of properly, or if septic tanks leak, it can contaminate rivers, ponds, and the ocean. Businesses and homeowners cannot claim that waste came from other animals when DNA evidence can prove that it came from humans. Polluters can then be forced to clean up the environment.

## HOW ARE WE RELATED?

Scientists are also using DNA forensics to trace family ancestries in humans, plants, and animals. Once genomes, or even parts of them, have been sequenced, they can be compared to see how closely related different organisms, or even individuals, are. This is particularly useful when two species might not look at all alike on the outside.

The hyrax is a small, furry animal that lives in rock piles in Africa. The African elephant is a gigantic animal that lives in grasslands. DNA analysis has shown that they are more closely related to each other than to any other animal. Scientists have also used DNA from the frozen cells of a wooly mammoth to not only prove they were related to elephants, but that they were more closely related to Asian elephants than African elephants.

DNA fingerprinting could help ranchers maintain the bloodlines of valuable livestock, such as cattle, sheep, or horses. Breeders of any type of plant or animal would know what genes their breeding stock carried. They could avoid breeding individuals that were shown to carry defective or lethal genes. Captive breeding programs of wild animals in zoos and parks would also benefit from DNA analysis.

Tampering with nature can backfire. Once modified genes become part of an organism's DNA, they are there forever. Scientists worry that mutations in modified genes and viruses could cause side effects or diseases worse than the original problems they were used to treat. Modified organisms could escape into the wild and alter wild populations forever.

Now that the human genome has been cracked, scientists can use DNA fingerprints to trace family histories. Relatives separated for many years by disaster or war could be reunited. Human migration patterns around the world could be followed. Perhaps one day you will have a family album—a really big one—that shows how closely related you are to every other person on Earth! Or a family tree that traces your ancestors back thousands of years. Who knows, you might find you are related to Cleopatra, King Tut, or Alexander the Great.

Scientists have already used DNA fingerprinting to prove the close relationship between humans and the extinct Neanderthal, a primitive human that lived in Europe during the Stone Age. They used DNA from a Neanderthal bone that was estimated to be about 38,000 years old, from a time when it is believed that Neanderthals and humans lived in close proximity. So far they have sequenced 1 million base pairs from the Neanderthal DNA, out of a total of 3 billion base pairs—the same number as found in humans. The results showed that Neanderthals and humans share at least 99.5 percent of the same DNA.

And straight from the pages of Michael Crichton's *Jurassic Park*, scientists are retrieving DNA from fossil insects, including termites and bees, and bacteria that have been preserved for millions of years in amber. Amber is the petrified sap from ancient trees. The oldest known fossil ever found was a bacterium from the stomach

An insect preserved in amber may still carry intact DNA from millions of years ago.

of a preserved bee that lived about 100 million years ago. By sequencing the DNA in fossils, scientists can learn about the past and how it relates to the present.

Scientists have even tried to bring back the mammoth by placing its DNA into the egg of an elephant. So far they have not succeeded—and some people think they never should—but it may not be long before someone succeeds. Dinosaur zoos could be a part of your future after all. And as scientists continue on their quest to recreate life, your next-door neighbor might someday be a Neanderthal! We lived together once; why not again?

# Glossary

**adapt**—to become fit for survival in an environment through changes in the DNA

**anticodon**—a nucleotide triplet on a tRNA molecule that is complementary to the codon triplet on an mRNA molecule

**asexual reproduction**—the production of offspring involving only one parent in which an identical set of the parent's DNA is passed on

**bacterium** (*plural*, bacteria)—an extremely small, single-celled organism without a nucleus; a prokaryote

**base**—a chemical in a DNA or RNA molecule that, when repeated, carries gene messages

**bioremediation**—the use of biological organisms to clean up pollutants in the environment

**botanist**—a scientist who studies plants

**cancer**—uncontrolled cell growth

**chromatin**—long strand of DNA and protein that carries the genes

**chromosomes**—the rod-like bodies of tightly-coiled chromatin that are seen during cell division

**classify**—to organize into groups based on similar characteristics

**codon**—a triplet of bases in a DNA or RNA molecule that specifies a single amino acid when proteins are made

**complementary bases**—bases on the DNA and RNA that chemically bond to one another

**complex**—composed of many parts

**data**—factual information used as a basis for reasoning

**deoxyribonucleic acid**—a nucleic acid that controls the activities of all living cells; DNA

**descendant**—organism that comes from another organism or from an ancestor

**diffraction**—the breaking up of light into the colors of the spectrum

**dominant trait**—an inherited characteristic that is expressed or appears in an individual

**double helix**—the twisted arrangement of DNA made from two strands held together by bases

**electron**—a subatomic particle with a negative electrical charge

**embryo**—an organism in the earliest stage of its development

**enzyme**—a protein that speeds up chemical reactions inside of a cell

**essential amino acid**—an amino acid that is not produced by cells and must be acquired in the diet

**eukaryote**—a cell that has a nucleus and other membrane-covered organelles

**excrete**—to eliminate or remove wastes

**extract**—a product that is removed or taken out of something else

**fertilize**—to cause an egg to begin development by uniting it with a sperm

**forensics**—the practice of medicine to help solve crimes; forensic medicine

**gene**—a segment of DNA on a chromosome that codes for a specific trait

**generation**—the time between the birth of a parent and the birth of that parent's offspring

**genetic disorder**—a condition or disease caused by a problem or change in one or more genes

**genetic engineering**—the manipulation of a gene or genes in an organism with the aim of introducing new characteristics

**genome**—all of the genes found in one set of an organism's chromosomes

**genotype**—all or part of the genetic composition of an organism or a group of organisms

**hemoglobin**—a protein that binds to oxygen on red blood cells

**heredity**—the passing of traits from parents to offspring through DNA

**hypothesis**—a possible explanation or answer to a question; an educated guess

**immune system**—a group of organs that work together to help your body fight disease; lymphatic system

**inherit**—to receive genes from an ancestor

**interphase**—the stage in a cell's life when the cell grows and copies its DNA; the stage before cell division

**keratin**—a water-resistant protein found in skin, hair, and nails

**ligament**—a band of tissue that connects two bones together

**marsupial**—a mammal whose young are born live at an early stage of development and complete their development inside of a pouch on the mother's body

**micrometer**—a unit of length equal to one-millionth of a meter; also called a micron

**mitochondria** (*singular*, mitochondrion)—organelles in a cell that break down food to make energy for cell activities

**molecule**—a particle made of two or more atoms joined by chemical bonds

**multicellular**—made of many cells

**mutagen**—any agent that causes a change in the base sequence of a DNA molecule

**mutant**—a cell or organism carrying a gene that has been altered or changed

**mutation**—an alteration or change in the order of the bases in an organism's DNA

**nucleic acid**—a very large molecule made of chemicals that hold the instructions cells need to carry out the activities of life

**nucleosome**—a subunit of chromatin consisting of a length of DNA wound around a cluster of proteins

**nucleotide**—the building block of a nucleic acid consisting of a phosphate group, a sugar, and a base

**nucleus**—the organelle in a eukaryotic cell that holds the DNA

**offspring**—the young of an organism

**organ**—a structure made of different kinds of tissue that work together to do a particular job

**organelle**—a structure inside of a cell that carries out a specific job for the cell

**organism**—a living thing made of one or more cells that contain DNA

**pathogen**—an organism or virus that causes disease

**phenotype**—the inherited appearance of an organism

**plankton**—microscopic organisms that float near the surface of warm bodies of water

**plasmid**—a circular piece of DNA outside of the normal chromosome of a bacterium that sometimes contains genes responsible for antibiotic resistance

**prokaryote**—a cell that lacks a nucleus and membrane-covered organelles; a bacterium

**protein**—a large molecule needed for growth, repair, and chemical reactions in a cell

**recessive trait**—a characteristic that is hidden when it is paired with the dominant trait

**recombinant DNA**—a DNA molecule formed by the joining of genes from two different organisms

**redundant**—repetitive; in DNA or RNA, when more than one codon means the same thing

**replicate**—to undergo replication

**replication**—the process of synthesizing or copying DNA

**ribonucleic acid**—a nucleic acid containing a ribose sugar that acts in making proteins; RNA

**ribosome**—a small organelle in the cell where proteins are made

**self-pollinate**—to transfer pollen from the male part of one flower to the female part of the same flower, or to a flower on the same plant

**sex chromosome**—a chromosome that carries the gene to determine the gender of an individual

**sexual reproduction**—reproduction involving two parents who each contribute half their genes to produce a new individual

**species**—a group of similar organisms that can mate with each other and produce fertile offspring

**stimulus**—anything that causes a response or change in a cell, tissue, organ, or organism

**strain**—one of a group of organisms that come from a common ancestor

**structure**—the arrangement of the parts that make up an item

**synthetic**—produced artificially rather than occurring naturally

**tendon**—a connective tissue that connects skeletal muscle to bone

**tissue**—a group of similar cells that work together to do a particular job

**transcription**—the transfer of genetic information from DNA to messenger RNA

**translation**—the building of a protein at the site of a ribosome according to the codons on a messenger RNA molecule

**triplet**—a set of three bases that passes the genetic code from DNA to RNA

**unicellular**—made of a single cell

**virus**—an extremely small molecule of DNA and protein, which can only reproduce by invading a living cell

**x-ray diffraction**—a technique used to determine the structure of molecules through patterns that are made by the scattering of x-rays through crystals

# Search Engine

## BOOKS

Campbell, Neil A., Lawrence G. Miller, and Jane B. Reece. *Biology: Concepts and Connections.* Menlo Park, CA: Benjamin Cummings, 1998.

Marieb, Elaine N., ed. *Human Anatomy and Physiology.* Redwood City, CA: Benjamin Cummings, 1995.

Mattox, Brenda. *Rosalind Franklin: The Dark Lady of DNA.* New York: HarperCollins, 2002.

*McGraw-Hill Encyclopedia of Science and Technology.* 8th ed. New York: McGraw-Hill, 1997.

Snedden, Robert. *Cell Division and Genetics.* Chicago: Heinemann Library, 2003.

————. *DNA and Genetic Engineering.* Chicago: Heinemann Library, 2003.

## WEB SITES

BBC News Online
   Speech genes (Briggs, Helen. *First Language Gene Discovered.* BBC News Online, 2002)
   www.news.bbc.co.uk

Bionet
   Genetically modified crops
   www.bionetonline.org/English/Content/ff_intro.htm

Cells Alive!
   Interactive tour of the cell
   www.cellsalive.com

ChemforKids
   Basics of biochemistry, including DNA facts
   www.chemforkids.com

Dolan DNA Learning Center
   www.dnalc.org

Genome News Network
   www.genomenewsnetwork.org

Genome Sequencing Center
   genome.wustl.edu

Human Genome Project
   www.genome.gov

National Geographic
   Information on the DNA computer
   www.nationalgeographic.com
National Health Museum
   DNA vaccines
   www.accessexcellence.org
Nature Journal
   Evolutionary genetics and changes in chromosome number
   www.nature.com
NewScientist
   DNA vaccine for tuberculosis
   www.newscientist.com
Oak Ridge National Laboratory
   Human genome information
   www.ornl.gov
Proceedings of the National Academy of Sciences
   www.pnas.org
Royal Society of New Zealand
   Genes and biology
   www.rsnz.org
Think Quest
   Links to student-created Web sites on DNA and genes
   www.thinkquest.org
U.K. Cancer Research
   info.cancerresearchuk.org
University of Utah Genetic Science Learning Center
   learn.genetics.utah.edu
U.S. News and World Report
   General articles about DNA
   www.usnews.healthline.com
USA Today News
   Neanderthal genes (Vergano, Dan. *Scientists Mapping Genetic Blueprint of
      Neanderthals*. USA Today Online, 2006).
   www.usatoday.com/tech/science/discoveries
Your Genes, Your Health
   Genetic disorders
   www.ygyh.org

# Index

*Page numbers in italics refer to illustrations.*

# About the Author

**Susan Schafer** is a science teacher and the author of several nonfiction books for children. She has written about numerous animals, including horses, snakes, tigers, Komodo dragons, and Galapagos tortoises. Her book on the latter was named an Outstanding Science Trade Book for Children by the National Science Teachers Association and Children's Book Council. She has also written a fictional book about animal tails for very young children. Schafer has spent many years working in the field of biology and enjoys sharing her knowledge and appreciation of nature with others. She lives on a ranch in Santa Margarita, California, with her husband, horses, and dogs, and with the beauty of the oak-covered hills around her.

T - #0922 - 101024 - C96 - 246/189/4 - PB - 9780765683076 - Gloss Lamination